A Panoply of Polygons

AMS / MAA | DOLCIANI MATHEMATICAL EXPOSITIONS

VOL 58

A Panoply of Polygons

Claudi Alsina
Roger B. Nelsen

Providence, Rhode Island

Dolciani Mathematical Expositions Editorial Board
C. Ray Rosentrater, Editor

Priscilla S. Bremser   Heather A. Dye   Kim Regnier Jongerius
Alfred M. Dahma   C. L. Frenzen   Katharine Ott
Matt Delong   Jennifer R. Galovich   Tom Richmond

2020 *Mathematics Subject Classification.* Primary 51M05, 51M15, 97G40.

For additional information and updates on this book, visit
www.ams.org/bookpages/dol-58

Library of Congress Cataloging-in-Publication Data

Names: Alsina, Claudi, author. | Nelsen, Roger B., author.
Title: A panoply of polygons / Claudi Alsina, Roger B. Nelsen.
Description: Providence, Rhode Island : MAA Press, an imprint of the American Mathematical Society, [2023] | Series: Dolciani mathematical expositions ; volume 58 | Includes bibliographical references and index.
Identifiers: LCCN 2022046870 | ISBN 9781470471842 (paperback) | ISBN 9781470472948 (pdf)
Subjects: LCSH: Polygons. | AMS: Geometry – Real and complex geometry – Euclidean geometries (general) and generalizations. | Geometry – Real and complex geometry – Geometric constructions. | Mathematics education – Geometry – Plane and solid geometry.
Classification: LCC QA482 .A476 2023 | DDC 516/.154–dc23/eng20230112
LC record available at https://lccn.loc.gov/2022046870

Copying and reprinting. Individual readers of this publication, and nonprofit libraries acting for them, are permitted to make fair use of the material, such as to copy select pages for use in teaching or research. Permission is granted to quote brief passages from this publication in reviews, provided the customary acknowledgment of the source is given.

Republication, systematic copying, or multiple reproduction of any material in this publication is permitted only under license from the American Mathematical Society. Requests for permission to reuse portions of AMS publication content are handled by the Copyright Clearance Center. For more information, please visit www.ams.org/publications/pubpermissions.

Send requests for translation rights and licensed reprints to reprint-permission@ams.org.

© 2023 by the American Mathematical Society. All rights reserved.
The American Mathematical Society retains all rights
except those granted to the United States Government.
Printed in the United States of America.

∞ The paper used in this book is acid-free and falls within the guidelines
established to ensure permanence and durability.
Visit the AMS home page at https://www.ams.org/

10 9 8 7 6 5 4 3 2 1   28 27 26 25 24 23

Contents

Preface ix

1 Polygon Basics 1
   1.1 Introduction 1
   1.2 Polygon preliminaries 2
   1.3 Diagonals of convex polygons 5
   1.4 Regular polygons 7
   1.5 Drawing regular polygons and the Gauss-Wantzel theorem 14
   1.6 Star polygons 16
   1.7 Polygonal tiling 19
   1.8 Voronoi diagrams and dual tilings 23
   1.9 Challenges 25

2 Pentagons 29
   2.1 Introduction 29
   2.2 Regular pentagons 30
   2.3 Drawing a regular pentagon 35
   2.4 Folding a regular pentagon 38
   2.5 Six types of general convex pentagons 39
   2.6 Pentagonal tilings 43
   2.7 Pentagrams 48
   2.8 Pentagons in space 52
   2.9 Miscellaneous examples 54
   2.10 Pentagons in architecture 56
   2.11 Challenges 57

3 Hexagons — 61

- 3.1 Introduction — 61
- 3.2 Regular hexagons — 63
- 3.3 Cyclic hexagons — 71
- 3.4 Hexagonal tilings — 75
- 3.5 Parahexagons — 77
- 3.6 The carpenter's square — 81
- 3.7 L-polyominoes — 82
- 3.8 Hexagrams — 86
- 3.9 Miscellaneous examples — 90
- 3.10 Hexagons in architecture — 95
- 3.11 Challenges — 97

4 Heptagons — 103

- 4.1 Introduction — 103
- 4.2 Regular heptagons — 105
- 4.3 The diagonals of a regular heptagon — 105
- 4.4 The heptagonal triangle — 107
- 4.5 Drawing a regular heptagon — 109
- 4.6 A neusis construction — 112
- 4.7 Heptagonal tilings — 114
- 4.8 Star heptagons — 115
- 4.9 Heptagons in architecture — 116
- 4.10 Challenges — 118

5 Octagons — 121

- 5.1 Introduction — 121
- 5.2 Regular octagons — 122
- 5.3 General convex octagons — 128
- 5.4 Star octagons — 131
- 5.5 Octagons in space — 133
- 5.6 Octagons in architecture — 134

| | 5.7 | Challenges | 137 |

6 Many-sided Polygons — 141

- 6.1 Introduction — 141
- 6.2 Nonagons — 143
- 6.3 Decagons — 148
- 6.4 Hendecagons — 153
- 6.5 Dodecagons — 155
- 6.6 Gauss and heptadecagons — 161
- 6.7 Archimedes and 24-, 48-, and 96-gons — 162
- 6.8 The 257-gons and the 65537-gons — 166
- 6.9 Miscellaneous many-sided polygons — 167
- 6.10 Challenges — 170

7 Miscellaneous Classes of Polygons — 175

- 7.1 Introduction — 175
- 7.2 Lattice polygons — 175
- 7.3 Rectilinear polygons — 181
- 7.4 Zonogons — 183
- 7.5 Cyclic polygons — 186
- 7.6 Tangential polygons — 190
- 7.7 Bicentric polygons — 194
- 7.8 Challenges — 195

8 Polygonal Numbers — 197

- 8.1 Introduction — 197
- 8.2 Ordinary polygonal numbers — 200
- 8.3 Centered polygonal numbers — 205
- 8.4 Other figurate numbers derived from polygons — 207
- 8.5 Challenges — 209

Solutions to the Challenges 213
    Chapter 1 213
    Chapter 2 215
    Chapter 3 219
    Chapter 4 226
    Chapter 5 230
    Chapter 6 234
    Chapter 7 240
    Chapter 8 242

Credits and Permissions 247

Bibliography 253

Index 261

Preface

Geometry is the archetype of the beauty of the world.
Johannes Kepler

Geometry is nothing at all, if not a branch of art.
Julian Lowell Coolidge

Beauty is geometry.
J. K. Rowling

To be a well-educated person, the study of geometry should be included in one's academic background. And some of the most enchanting objects in elementary geometry are the *polygons*, figures constructed from two basic geometric concepts, line segments and angles. They are prime examples of great beauty created from the simplest of mathematical objects, and are the subject of this Panoply.

What is a *Panoply*? Dictionary definitions include *a splendid or striking array or arrangement* and *a complete or impressive collection or display of something.* The objects in our Panoply are, for the most part, planar polygons with five or more sides. The restriction to five or more sides is because entire books can be and have been written about triangles (e.g., several books in Euclid's *Elements*) and quadrilaterals (e.g., the 2020 AMS-MAA book *A Cornucopia of Quadrilaterals*). But books devoted to polygons with five or more sides seem to be rather rare.

Polygons come in a variety of shapes and possess a multitude of different properties, as witnessed by the many adjectives used to describe them: *simple, complex, concave, convex, compound, equiangular, equilateral, regular, skew, star, cyclic, tangential, bicentric, rectilinear, lattice, equable*, etc. In addition we have the adjective *polygonal*, as in *polygonal numbers*, positive integers that enumerate sets of objects in polygonal patterns, numbers that have been studied since the time of the ancient Greek mathematicians. All of these types of polygons are present in our Panoply, and in many instances illustrated with examples from art and architecture.

A Panoply of Polygons consists of eight chapters. In the first chapter we present some basic facts about convex polygons and all regular polygons and regular star polygons, along with some mathematical tools applicable to polygons in subsequent chapters (e.g., Ptolemy's theorem to study cyclic polygons). In this chapter we also discuss the Gauss-Wantzel theorem and various methods for drawing regular polygons, and the regular polygons that appear in various regular and semiregular tilings of the plane.

Each of the next four chapters is devoted to a particular type of polygon. In Chapter 2 we present the properties of regular and irregular pentagons, including the relationship with the golden ratio, equidiagonal and parallel pentagons, pentagrams, and monohedral tilings of the plane with convex pentagons. Chapter 3 is devoted to hexagons, where we have not only regular hexagons and hexagrams, but also irregular ones such as parahexagons and certain polyominoes. In Chapter 4 we study heptagons and present some of the remarkable properties of the heptagonal triangle, whose sides are the two unequal diagonals and an edge of a regular heptagon. In addition we study the two regular star heptagons and present several ways to draw approximations of a regular heptagon. Chapter 5 presents octagons and octagrams in a manner analogous to the treatment of the polygons in the previous three chapters.

In Chapter 6 we have a selection of polygons with nine or more sides with some distinguishing properties, e.g., nonagons, decagons, and dodecagons. Others are included for their roles in the history of mathematics, such as the 17-gon and the 96-gon. Chapter 7 presents some classes of polygons sharing a common characteristic or property, such as lattice polygons, rectilinear polygons, and cyclic and tangential polygons. In Chapter 8 we conclude with a brief study of polygonal numbers, both ordinary and centered, and their relationships. Each chapter includes a set of exercises we call Challenges, with solutions to all the Challenges following Chapter 8. The book concludes with a bibliography and a complete index.

A Panoply of Polygons is not a textbook, although it can be used as a supplement to a high school or college geometry course. It can also be used as a source for group projects or extra-credit assignments. But more importantly, we believe it will be of interest to anyone who loves geometry.

A note about our notation: We use labels such as *AB* or *a* both for a line segment and for its length; the context suffices to indicate which. Similarly we use *A* or θ for both an angle and its measure (*A* may also be a vertex of a polygon). We use both degrees and radians for angle measurement; whichever is more convenient in a particular situation. The symbol ∎ marks the end of a proof, while □ marks the end of an example.

Acknowledgments. We would like to express our appreciation and gratitude to C. Ray Rosentrater and the members of the editorial board of the Dolciani series for their careful reading of earlier drafts of this book and their many helpful suggestions. We would also like to thank Stephen Kennedy and Carol Baxter of the MAA publication staff, and Christine Thivierge, Sergei Gelfand, and Thomas Costa of the AMS production staff for their encouragement, expertise, and hard work in preparing this book for publication.

Claudi Alsina
Universitat Politècnica de Catalunya
Barcelona, Spain

Roger B. Nelsen
Lewis & Clark College
Portland, Oregon

CHAPTER 1

Polygon Basics

Polygons are to planar geometry as integers are to numerical mathematics.

S. L. Devadoss and J. O'Rourke
Discrete and Computational Geometry

1.1. Introduction

Polygons have a role in mathematics that extends far beyond plane geometry. Since the time of the ancient Greek mathematicians, polygons have been employed to expand our knowledge in many fields of mathematics. For example, it was by considering the diagonals of a square and a pentagon that the Pythagoreans showed that $\sqrt{2}$ and the golden ratio were irrational numbers. Polygonal approximations to circles also yielded a variety of approximations to the number π. Today polygons are used outside of mathematics in fields such as organic chemistry, engineering, architecture, computer science and its applications, art and sculpture, etc.

This chapter presents some basic concepts and results that will be useful in the chapters to follow where we study specific polygons in more detail.

Polygons in Moorish art in Andalucía

When the Moors were in the southern part of Spain now known as Andalucía, they built beautiful palaces often decorated with geometric mosaics featuring a variety of polygons. The examples in Figure 1.1.1 are from the Reales Alcázares in Seville. On the left the design includes star octagons (see Chapter 5) and parahexagons (see Chapter 3); and on the right we have star hexagons (see Chapter 3) and a star dodecagon (see Chapter 6).

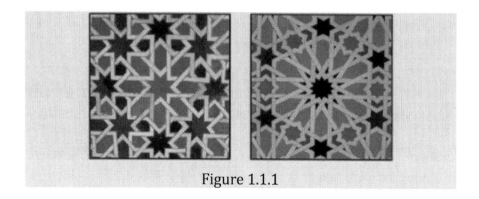

Figure 1.1.1

1.2. Polygon preliminaries

In this section we present some vocabulary that we employ in the chapters that follow, beginning with a few important definitions. To assist the reader, a sentence in this chapter with a term in **bold italics** indicates such a definition.

- A ***polygon*** is a plane figure consisting of n distinct points $v_1, v_2, v_3, \ldots, v_n$ called *vertices*, with no three successive points collinear, and n line segments $v_1 v_2, v_2 v_3, v_3 v_4, \ldots, v_n v_1$ called *sides* or edges, for $n \geq 3$ [Coxeter and Greitzer, 1967; James and James, 1992]. A polygon with n sides is often called an ***n-gon***. This definition of a polygon yields a ***unicursal*** figure, one that can be drawn with line segments vertex to vertex without lifting the pen from the paper.

- A polygon is ***simple*** if no two sides cross one another. The plane region enclosed by the sides of a simple polygon is its ***interior***, and the (interior) ***angles*** of a polygon are the angles made by adjacent sides and lying in the interior.

- A simple polygon is ***convex*** if it lies on one side of any line containing a side of the polygon. The interior angles of a convex polygon all measure less than 180°.

- A simple polygon is ***concave*** if it is not convex, i.e., if and only if there is a line that passes through the interior and cuts the polygon in four or more points. At least one interior angle of a concave polygon is a **reflex angle**, one that measures more than 180°

• A polygon is **complex** if it is not simple, i.e., the sides in at least one pair cross one another.

• A polygon is **compound** if it is a figure that is the union of two or more simple polygons.

See Figure 1.2.1 for examples of some of the polygons defined above.

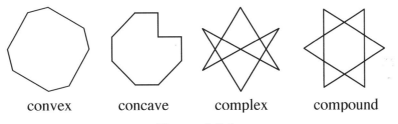

convex   concave   complex   compound

Figure 1.2.1

Our definition of a polygon does not include the one in Figure 1.2.1 labeled "compound" as it is not unicursal. It is an example of a **star polygon**, which we define and discuss in Section 1.6. There are other unions of two simple polygons that we will not consider to be compound polygons in this book. For example, the unions of two triangles (six vertices, six sides, but not a simple hexagon nor a star hexagon) in Figure 1.2.2 are not polygons. See [Malkevitch, 2016] for an essay about the difficulty of giving a precise definition of a polygon.

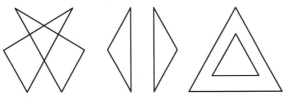

Figure 1.2.2

• A **skew polygon** is one whose vertices do not all lie in the same plane (note that our definition of *polygon* states that it is planar). Simple examples include ones constructed from the edges of a polyhedron. In Figure 1.2.3 we see a pentagonal antiprism, consisting of two regular pentagons in parallel planes and ten equilateral triangles. The ten edges that connect one pentagon to the other form an equilateral equiangular skew decagon (a 10-gon).

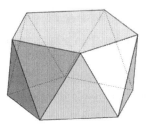

Figure 1.2.3

- A polygon is **cyclic** if it possesses a **circumcircle** (with its **circumcenter** and **circumradius**), a circle passing through each vertex; and **tangential** if it possesses an **incircle** (with its **incenter** and **inradius**), a circle tangent to each side. See Figure 1.2.4. We study a variety of properties of simple cyclic and tangential polygons in Chapter 7. A **bicentric** polygon is one that is both cyclic and tangential.

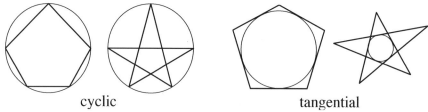

cyclic   tangential

Figure 1.2.4

- A polygon is **equiangular** if its interior angles have equal measure, and **equilateral** if its sides all have the same length. A **regular** polygon is one that is both equiangular and equilateral. In Section 1.4 we show that regular polygons are bicentric. However, a bicentric polygon need not be regular, as shown by the irregular bicentric pentagon in Figure 1.2.5.

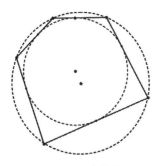

Figure 1.2.5

1.3. Diagonals of convex polygons

A *diagonal* of a polygon is a line segment joining two nonadjacent vertices. In the next two theorems we present basic properties of the diagonals in convex polygons.

Theorem 1.3.1. *In a convex n-gon we have*

(i) *the sum of the vertex angles is $(n-2)\pi$ (or $(n-2)180°$),*

(ii) *there are $n(n-3)/2$ diagonals, and*

(iii) *the diagonals intersect in at most $\binom{n}{4}$ interior points.*

Proof. Each vertex of an n-gon is connected to two adjacent vertices by sides of the n-gon, and to the remaining $n-3$ vertices by diagonals. Those $n-3$ diagonals partition the n-gon into $n-2$ triangles. Hence the sum of the interior angles of the n-gon is $(n-2)\pi$. For (ii) we note that $n-3$ diagonals terminate at each vertex, so that the number of endpoints of diagonals is $n(n-3)$. Since each diagonal has two endpoints, there are $n(n-3)/2$ diagonals. For (iii), each interior point of intersection of two diagonals is also the point of intersection of the diagonals of at least one quadrilateral whose four vertices are vertices of the n-gon (see Figure 1.3.1), and four vertices of the n-gon can be chosen in $\binom{n}{4}$ ways. ∎

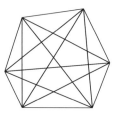

Figure 1.3.1

Now let $I(n)$ be the number of intersection points inside a *regular n-gon* formed by its diagonals. By Theorem 1.2.1(iii) we have $I(n) \leq \binom{n}{4}$ for all n. However, it can be shown that for n odd there are no intersections of three or more diagonals [Heineken, 1962], so that $I(n) = \binom{n}{4}$ for odd n. When n is even and at least 6, $I(n) < \binom{n}{4}$ since such an n-gon will have at least three diagonals (diameters of the circumcircle) intersecting at the center. The exact formula for $I(n)$ in this case is rather complicated, see [Poonen and Rubenstein, 1998] for details.

We now count the number of triangles in the interior of an n-gon formed by its sides and diagonals. We consider the case where no three diagonals intersect at an interior point.

Theorem 1.3.2. *Let P be a convex n-gon with the property that no three diagonals are concurrent. Then the number Δ_n of triangles in P whose vertices are either interior points or vertices of P is*

$$\Delta_n = \binom{n}{3} + 4\binom{n}{4} + 5\binom{n}{5} + \binom{n}{6}.$$

Proof [Conway and Guy, 1996]. We count the triangles according to the number of vertices the triangle shares with P. There are $\binom{n}{3}$ triangles with all three vertices in common with P (see Figure 1.3.2a), $4\binom{n}{4}$ triangles that have exactly two vertices in common with P (see Figure 1.3.2b), $5\binom{n}{5}$ triangles that have exactly one vertex in common with P (see Figure 1.3.2c), and $\binom{n}{6}$

triangles all of whose vertices are interior points of *P* (see Figure 1.3.2d). ∎

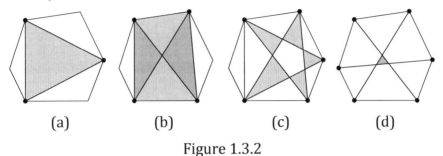

(a) (b) (c) (d)

Figure 1.3.2

With standard binomial coefficient identities the expression in the above theorem simplifies to $\Delta_n = \binom{n+3}{6} + \binom{n+1}{5} + \binom{n}{5}$. Elements of the sequence $\{\Delta_n\}_{n=3}^{\infty}$ appear as sequence A005732 in the *Online Encyclopedia of Integer Sequences* at oeis.org.

In the next section we present a variety of observations about regular polygons. One reason that regular polygons are important and interesting is described in the following theorem.

Theorem 1.3.3 The isoperimetric theorem for polygons. *Among all n-gons with a given perimeter P, the regular n-gon has the greatest area.*

See [Demar, 1975] for a proof showing that the *n*-gon with maximum area must be both equilateral and equiangular.

We note here that it is easy to "square" any simple polygon, i.e., to draw a square with the same area as the polygon. See Proposition 14 in Book II of Euclid's *Elements*.

1.4. Regular polygons

A central topic in the chapters to follow is the regular polygon with *n* sides (or *n*-gon) of length *a*, and *n* vertex angles with measure θ (in either degrees or radians). A regular *n*-gon is a union of *n* isosceles triangles with two base angles each with measure $\theta/2$ and an angle between the two legs with measure $2\pi/n$. Thus $\theta = (n-2)\pi/n$. The vertex angles meet at a point O that is equidistant from the vertices of the n-gon, so that a regular polygon possesses a circumcircle and is cyclic with circumcenter O and circumradius R, the length of the legs of the aforementioned isosceles triangles. See Figure 1.4.1.

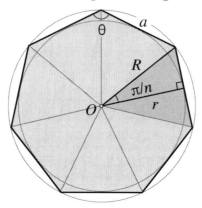

Figure 1.4.1

The point O is also equidistant from the midpoint of each side, so that a regular polygon possesses an incircle and is tangential with inradius r (and whose incenter is also O). We also denote the area of the regular n-gon by K. Then elementary geometry and trigonometry yield the following data for a regular polygon with n sides.

vertex angle	$\theta = \dfrac{n-2}{n}\pi$ or $\dfrac{n-2}{n}180°$
side length	$a = 2r\tan\dfrac{\pi}{n} = 2R\sin\dfrac{\pi}{n}$
semiperimeter	$s = an/2$
area	$K = \dfrac{n}{4}a^2\cot\dfrac{\pi}{n} = nr^2\tan\dfrac{\pi}{n} = \dfrac{n}{2}R^2\sin\dfrac{2\pi}{n} = rs$
circumradius	$R = \dfrac{a}{2}\csc\dfrac{\pi}{n}$
inradius	$r = \dfrac{a}{2}\cot\dfrac{\pi}{n} = R\cos\dfrac{\pi}{n}$

1.4. REGULAR POLYGONS

Example 1.4.1. *The isoperimetric inequality for polygons.* At the end of the previous section we presented the isoperimetric theorem for *n*-gons. From the data above, the area of a regular *n*-gon in terms of its perimeter P is $(P^2/4n)\cot(\pi/n)$. Hence if K and P are the area and perimeter, respectively, of an *n*-gon, then the isoperimetric theorem yields the *isoperimetric inequality for n-gons*: $4nK \leq P^2 \cot(\pi/n)$ with equality if and only if the *n*-gon is regular. □

In the following theorem we present several relationships for the perimeters and areas of regular *n*-gons and regular 2*n*-gons inscribed in and circumscribed about the same circle. The proof employs the above data for regular *n*-gons. Relationships such as these enabled Archimedes to derive the inequality $3\frac{10}{71} < \pi < 3\frac{1}{7}$ by considering 96-gons inscribed in and circumscribed about a unit circle (a circle with radius 1). For details see Section 6.7.

Theorem 1.4.1. *Let p_n and k_n denote, respectively, the perimeter and area of a regular n-gon inscribed in a circle of radius ρ, and let P_n and K_n denote, respectively, the perimeter and area of a regular n-gon circumscribed about the same circle. Then*

(1.1) $$\rho p_n = 2k_{2n} \quad \text{and} \quad \rho P_n = 2K_n,$$

(1.2) $$P_{2n} = \frac{2p_n P_n}{p_n + P_n} \quad \text{and} \quad p_{2n} = \sqrt{p_n P_{2n}},$$

(1.3) $$k_{2n} = \sqrt{k_n K_n} \quad \text{and} \quad K_{2n} = \frac{2k_{2n}K_n}{k_{2n} + K_n}.$$

Proof. First note that ρ is the circumradius of the inscribed polygons and the inradius of the circumscribed polygons. Let $\alpha_n = \pi/n$. Hence

$$p_n = 2n\rho \sin \alpha_n \quad \text{and} \quad P_n = 2n\rho \tan \alpha_n,$$

and

$$k_n = \frac{n}{2}\rho^2 \sin 2\alpha_n = n\rho^2 \sin \alpha_n \cos \alpha_n \quad \text{and} \quad K_n = n\rho^2 \tan \alpha_n.$$

For (1.1) we have $\rho p_n = 2n\rho^2 \sin \alpha_n = 2k_{2n}$ (since $\alpha_n = 2\alpha_{2n}$); and $\rho P_n = 2K_n$ is twice the statement that the area (K_n) of a regular polygon equals its inradius (ρ) times its semiperimeter ($P_n/2$).

For (1.2) we have

$$\frac{2p_n P_n}{p_n + P_n} = \rho \frac{8n^2 \sin \alpha_n \tan \alpha_n}{2n(\sin \alpha_n + \tan \alpha_n)} = 4n\rho \frac{\sin \alpha_n}{1+\cos \alpha_n}$$
$$= 4n\rho \tan \alpha_{2n} = P_{2n},$$

and

$$p_n P_{2n} = 8(n\rho)^2 \frac{\sin^2 \alpha_n}{1+\cos \alpha_n} = 16(n\rho)^2 \frac{1-\cos \alpha_n}{2}$$
$$= (4n\rho)^2 \sin^2 \alpha_{2n} = p_{2n}^2.$$

For (1.3) we have, using both (1.1) and (1.2),

$$k_n K_n = (n\rho^2)^2 \sin^2 \alpha_n = (n\rho^2 \sin 2\alpha_{2n})^2 = k_{2n}^2,$$

and

$$\frac{2k_{2n} K_n}{k_{2n} + K_n} = \frac{\rho p_n \cdot \rho P_n / 2}{\rho p_n / 2 + \rho P_n / 2} = \rho \frac{p_n P_n}{p_n + P_n} = \rho P_{2n}/2 = K_{2n}. \blacksquare$$

The expressions in (1.2) and (1.3) for the perimeters and areas of the $2n$-gons are well-known *means* of two positive numbers: if x and y are positive numbers, then the *geometric mean* of x and y is $\sqrt{xy}$, and the *harmonic mean* is $2xy/(x+y)$.

The corresponding side lengths a_n and A_n of n-gons inscribed in and circumscribed about the same circle satisfy relationships similar to (1.2); see Challenge 1.2 at the end of this chapter.

In the next theorem we show that the sum of the n distances from an arbitrary point inside a regular n-gon to its sides is a constant. It is a generalization of a theorem for equilateral triangles named for the Italian mathematician Vincenzo Viviani (1622–1703).

Viviani's Theorem 1.4.2. *For a regular n-gon, the sum of the perpendicular distances from an arbitrary interior point to the n sides equals n times the inradius.*

Proof. Let $h_1, h_2, h_3, \ldots, h_n$ denote the perpendicular distances from the given point to the n sides and r the inradius of the n-gon, as illustrated in Figure 1.3.2a. We must show that

$$h_1 + h_2 + h_3 + \cdots + h_n = nr.$$

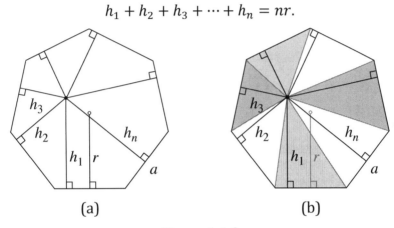

Figure 1.4.2

As before we let a denote the side length and $s = an/2$ the semiperimeter. Lines drawn from the arbitrary interior point to the vertices partition the n-gon into n triangles, as shown in Figure 1.3.2b. The areas of the triangles sum to the area K of the n-gon, so that we have

$$\frac{1}{2}ah_1 + \frac{1}{2}ah_2 + \frac{1}{2}ah_3 + \cdots + \frac{1}{2}ah_n = K = rs = r \cdot \frac{1}{2}an,$$

which proves the theorem. ∎

In several of the chapters to follow we consider complementary questions about sums of distances from a point on the circumcircle of a regular n-gon to its vertices.

As noted in Theorem 1.3.1 an n-gon has $n(n-3)/2$ diagonals, $n-3$ emanating from each vertex. In a regular n-gon those diagonals come in pairs, except for one that is a diameter of the circumcircle when n is even. We let d_k denote the length of a diagonal joining two vertices separated by k vertices, or equivalently, separated by $k+1$ sides of the n-gon. Hence $d_0 = a = 2R\sin(\pi/n)$ from the data on page 8 for a side of the n-gon, and when n is even, $d_{(n/2)-1} = 2R$, the length of a diameter of the n-gon. For the length d_k in general, see Figure 1.4.3.

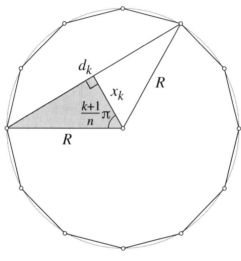

Figure 1.4.3

Let x_k be the length of the line segment joining the center of the circumcircle to the midpoint of one of the diagonals with length d_k, as shown in Figure 1.4.3. The measure of the angle between the two circumradii of length R is $(k+1)/n$ times 2π, so that the marked angle in the gray right triangle has measure $(k+1)/n$ times π. It now follows that

$$(1.4) \qquad d_k = 2R\sin\frac{k+1}{n}\pi \quad \text{and} \quad x_k = R\cos\frac{k+1}{n}\pi$$

for $1 \le k < n/2$. Since $R = \frac{a}{2}\csc\frac{\pi}{n}$ we have $d_k = a\sin\frac{(k+1)\pi}{n}\csc\frac{\pi}{n}$ for $n \ge 2(k+1)$. Thus $d_0 = a = 2R\sin(\pi/n)$ and $d_{(n/2)-1} = 2R$ for n even as noted above. Since $R = \frac{a}{2}\csc\frac{\pi}{n}$ we have $d_k = a\sin\frac{(k+1)\pi}{n}\csc\frac{\pi}{n}$ for $n \ge 2(k+1)$. Thus

$$(1.5) \begin{cases} d_1 = a \sin\dfrac{2\pi}{n} \csc\dfrac{\pi}{n} = 2a\cos\dfrac{\pi}{n} \text{ for } n \geq 4, \\ d_2 = a \sin\dfrac{3\pi}{n} \csc\dfrac{\pi}{n} = a\left(3 - 4\sin^2\dfrac{\pi}{n}\right) \\ = a\left(1 + 2\cos\dfrac{2\pi}{n}\right) \text{ for } n \geq 6, \\ d_3 = a \sin\dfrac{4\pi}{n} \csc\dfrac{\pi}{n} = 2a\sin\dfrac{2\pi}{n}\cos\dfrac{2\pi}{n}\csc\dfrac{\pi}{n} \\ = 4a\cos\dfrac{\pi}{n}\cos\dfrac{2\pi}{n} \text{ for } n \geq 8, \text{ etc.} \end{cases}$$

The above diagonals are the sides of the star polygons that we discuss in Section 1.6.

In the chapters that follow we derive a variety of relationships for the sides and diagonals of regular n-gons. Any three vertices of an n-gon form a triangle, for which we use tools such as the law of sines and the law of cosines. Any four vertices form a cyclic quadrilateral, for which our primary tool will be *Ptolemy's theorem*, named for Claudius Ptolemy of Alexandria (circa 85-165). Our proof is from Ptolemy's *Almagest*, but in modern notation.

Ptolemy's Theorem 1.4.3. *In a cyclic quadrilateral the product of the lengths of the diagonals is equal to the sum of the products of the lengths of the opposite sides.*

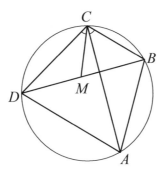

Figure 1.4.4

Proof. For the cyclic quadrilateral $ABCD$ in Figure 1.4.4, we need to show that

$$AC \cdot BD = AB \cdot CD + BC \cdot AD.$$

Let M be the point on diagonal BD such that $\angle DCM = \angle ACB$. From similar triangles DCM and ACB we have $CD/MD = AC/AB$ so that $AB \cdot CD = AC \cdot MD$. From similar triangles BCM and ACD we have $BC/BM = AC/AD$ so that $BC \cdot AD = AC \cdot BM$. It now follows that $AB \cdot CD + BC \cdot AD = AC(MD + BM) = AC \cdot BD$. ∎

Here is a typical application of Ptolemy's theorem relating the first two diagonals that applies to any regular n-gon with six or more sides.

Example 1.4.2. *Diagonals d_1 and d_2 in a regular n-gon.* Any four consecutive vertices of a regular n-gon with six or more sides of length a yield a cyclic quadrilateral with sides a, a, a, d_2 and diagonals d_1, d_2, where d_1 and d_2 are given by (1.5). By Ptolemy's theorem we have $d_1^2 = ad_2 + a^2$. □

1.5. Drawing regular pentagons and the Gauss-Wantzel theorem

Knowing how to draw a regular polygon is fundamental to the study of geometry. The first proposition in the first book of Euclid's *Elements*, i.e., Proposition I.1, reads as follows [Heath, 1956]: *On a given finite straight line to construct an equilateral triangle.* The tools permitted by Euclid in a construction are a compass and an unmarked straightedge (see Figure 1.5.1), which are the tools used to perform the tasks described in his first three postulates: to draw a straight line from any point to any point; to produce a finite straight line continuously in a straight line; and to describe a circle with any center and distance.

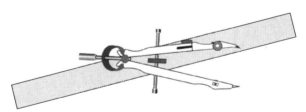

Figure 1.5.1

1.5. DRAWING REGULAR POLYGONS

In addition to the equilateral triangle, Euclid describes such constructions (called "classical" or "Euclidean" constructions) for a square in Proposition I.46, a regular pentagon in Proposition IV.11, and a regular hexagon in Proposition IV.15.

With Proposition III.30 (to bisect a given circumference) Euclid could construct a regular $2n$-gon from a regular n-gon, enabling the construction of regular octagons, decagons, dodecagons, etc. In Proposition IV.16 Euclid constructs a regular pentadecagon (a 15-gon) from an equilateral triangle and a regular pentagon, and this procedure enabled him to construct regular mn-gons from a regular m-gon and a regular n-gon when m and n are relatively prime, i.e., have no common divisors. So with results in the *Elements*, classical constructions of regular n-gons are possible for $n = 3, 4, 5, 6, 8, 10, 12, 15, 16, 20, 30, ...$, i.e., for n of the form $2^k m$ where k is a nonnegative integer and $m = 3, 4, 5$, or 15.

Is it possible to construct other regular n-gons with the Euclidean tools? That question remained unanswered for over 2000 years, until the following theorem was proved in the late 18th–early 19th century.

The Gauss-Wantzel Theorem 1.5.1. *A regular n-gon is constructible with compass and straightedge if and only if n is the product of a power of 2 and a nonnegative number of distinct Fermat primes, i.e., primes of the form* $2^{(2^k)} + 1$.

Gauss and Wantzel

The German mathematician Carl Friedrich Gauss (1777-1855) proved sufficiency (the "if" part of the theorem) in 1796, and the French mathematician Pierre Laurent Wantzel (1814-1848) proved necessity (the "only if" part) in 1837. See [Křížek et al., 2001] for a modern proof. For more details, see Chapter 6.

There are five known Fermat primes: 3, 5, 17, 257, and 65537; and it is conjectured that there are no more. Consequently, of the infinitely many values of n for which a regular n-gon is known to be constructible with straightedge and compass, only 31 are odd.

Thus for $n = 7, 9, 11, 13, 14, 18, 19, 21, 22, 23, \ldots$ no Euclidean construction of a regular n-gon is possible (see sequence A004169 in the *Online Encyclopedia of Integer Sequences* at oeis.org). For these n-gons there are two alternatives—use different tools (e.g., spirals, conic sections, computer graphics software, etc.) or settle for an approximate construction. If we replace the straightedge with a marked ruler we have a *neusis* construction, which we employ in Chapter 4 to draw a regular heptagon. A variety of approximate constructions of n-gons also appear in the following chapters. We note that pictures of most of the polygons, regular or otherwise, in this book were drawn with computer graphics software.

1.6. Star polygons

A ***star polygon*** is a regular complex or compound n-gon. Its sides are congruent diagonals of a regular n-gon. We employ the special notation $\{n/k\}$ for a star polygon with n vertices evenly spaced on a circle (i.e., the vertices of a regular n-gon) but whose sides join every kth vertex for $k < n/2$ (i.e., its sides are the diagonals d_{k-1}). In particular $\{n/1\} = \{n\}$, the regular n-gon. In a star polygon the sides cross one another, but the crossing points are not counted as vertices. In Figure 1.6.1 we see several examples of $\{n/k\}$ when $k \geq 2$.

1.6. STAR POLYGONS

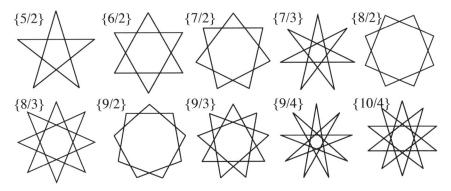

Figure 1.6.1

The number k in $\{n/k\}$ is the **density** of the star polygon. It gets that name from the fact that a ray from the circumcenter of $\{n/k\}$ not passing through a vertex crosses exactly k sides of the star polygon. We also note that the vertex angles of the star polygon $\{n/k\}$ measure $(n - 2k)\pi/n$.

When n and k are relatively prime, the star is unicursal. When n and k have a common divisor $d > 1$, then $\{n/k\}$ is compound and not unicursal, consisting of d copies of $\{\frac{n}{d}/\frac{k}{d}\}$, as in the cases $\{9/3\} = 3\{3\}$ and $\{10/4\} = 2\{5/2\}$ illustrated above. Some authors reserve the name "star polygon" for the relatively prime case and refer to $\{n/k\}$ when n and k have a common divisor $d > 1$ as a "star figure."

Interest in star polygons has a long history, dating back to the ancient Greek geometers, e.g., the pentagram $\{5/2\}$ was a symbol used by the Pythagoreans. The first persons to study star polygons in general were the English mathematician Thomas Bradwardine (1290-1349), the German mathematician Johannes Müller von Königsberg (1436-1476), better known as Regiomontanus, and the French mathematician Charles de Bouelles (1470-1566). The German astronomer and mathematician Johannes Kepler (1571-1630) discussed *stellated polygons* in his book *Harmonice Mundi* (Harmonies of the World) published in 1619.

Many objects called *stars* are polygons but not star polygons. For example, a five-pointed star such as one on the flag of the European Union or the United States is an equilateral concave decagon, with five 36° angles and five 252° reflex or reentrant angles, as shown on the left in Figure 1.6.2. Similarly, three-pointed and four-pointed stars are equilateral concave hexagons and octagons, respectively. On the right in Figure 1.6.2 we have an equilateral equiangular six-pointed star, but it is not a star polygon since its vertices are not vertices of a regular hexagon.

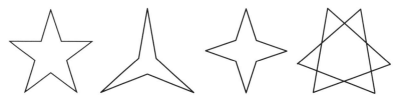

Figure 1.6.2

We define the area $[\{n/k\}, a]$ of the regular star polygon $\{n/k\}$ inscribed in a regular polygon with the same vertices and side length a as the area of the star shaded gray in Figure 1.6.3.

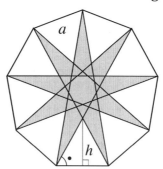

Figure 1.6.3

Theorem 1.6.1. *Let $[\{n/k\}, a]$ denote the area of a regular $\{n/k\}$ star polygon inscribed in a regular n-gon with the same vertices, side length a, and area K. Then*

(1.6) $$[\{n/k\}, a] = \frac{n}{4}a^2\left(\cot\frac{\pi}{n} - \tan\frac{(k-1)\pi}{n}\right).$$

When $k = 2$ we have

(1.7) $$[\{n/2\}, a] = \frac{n}{2}a^2 \cot\frac{2\pi}{n}.$$

Proof. In Figure 1.6.3 we have the gray star polygon $\{n/k\}$ inscribed in a regular n-gon with side a and area $K = (n/4)a^2 \cot(\pi/n)$. The area $[\{n/k\}, a]$ equals K minus n times the area of one of the white triangles with base a and height h.

The angle marked • equals $(k-1)\pi/n$ and thus $h = (a/2)\tan((k-1)\pi/n)$ and the area of one white triangle equals $(a^2/4)\tan((k-1)\pi/n)$, which yields the stated formula (1.6) for $[\{n/k\}, a]$. The special case (1.7) for $k = 2$ follows from the identity $\cot\theta - \tan\theta = 2\cot 2\theta$. ∎

We conclude this section with some data for a regular star polygon $\{n/k\}$ whose vertices are the vertices of a regular n-gon with side length a, diagonal d_{k-1}, and circumradius R. Note that the inradius of $\{n/k\}$ is the length x_{k-1} in Figure 1.4.3 and is given in (1.4).

vertex angle	$\theta = \frac{n-2k}{n}\pi$ or $\theta = \frac{n-2k}{n}180°$
side length	$d_{k-1} = a\sin\frac{k\pi}{n}\csc\frac{\pi}{n} = 2R\sin\frac{k\pi}{n}$
area	$[\{n/k\}, a] = \frac{n}{4}a^2\left(\cot\frac{\pi}{n} - \tan\frac{(k-1)\pi}{n}\right)$
circumradius	$R = \frac{a}{2}\csc\frac{\pi}{n}$
inradius	$x_{k-1} = R\cos\frac{k\pi}{n}$

1.7. Polygonal tiling

We begin this section with some useful vocabulary.

• A **tiling** (or **tessellation**) of the plane is any countable collection of closed sets (the *tiles*) that cover the plane without gaps or overlaps [Grünbaum and Shephard, 1987]. In this book the tiles are polygons.

- An **edge-to-edge tiling** is a polygonal tiling in which each edge of each polygon coincides with an edge of another polygon. A **vertex** of an edge-to-edge tiling is a vertex of one of the tiles.

- A **monohedral tiling** is one in which all the tiles are congruent, and a **uniform** tiling is one with identical vertices, i.e., one with the same arrangement of polygons at each vertex.

- A **regular tiling** is a monohedral uniform edge-to-edge tiling with regular polygons, and a **semiregular tiling** (or **Archimedean tiling**) is a uniform edge-to-edge tiling with two or more types of congruent regular polygons.

Example 1.7.1. *Polygonal floor tilings.* Regular and semiregular tilings have been employed for floor and wall coverings for centuries. In Figure 1.7.1a we see a Roman mosaic from the Museo Nazional di Palazzo Massimo alle Terme in Rome. It is a regular tiling with hexagons in which each hexagon has been partitioned into three congruent rhombi meeting at the hexagon's center, so that the tiling appears to be a collection of cubes.

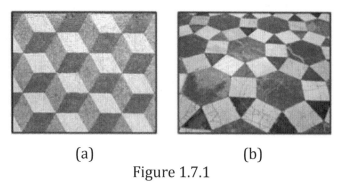

(a) (b)
Figure 1.7.1

In Figure 1.7.1b we have a semiregular tiling with triangles, squares and hexagons from the Museo Arqueológico de Sevilla in Spain. Note that each hexagon is surrounded by six triangles and six squares, forming a regular dodecagon (a 12-gon). □

1.7. POLYGONAL TILING

The above example raises the following questions: Which polygons form regular tilings? Which form semiregular tilings? The following two theorems answer those questions.

Theorem 1.7.1. *The only regular tilings consist of triangles, squares, and hexagons.* See Figure 1.7.2.

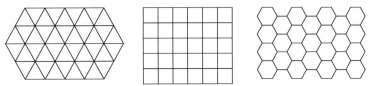

Figure 1.7.2

Proof. Suppose k regular n-gons meet at a vertex of the tiling. Since each vertex of a regular n-gon measures $(n-2)\pi/n$ we have $k(n-2)\pi/n = 2\pi$ or $(k-2)(n-2) = 4$. The only solutions to this equation in integers are $(k,n) = (3,6), (4,4)$, and $(6,3)$. ∎

Theorem 1.7.2. *The only semiregular tilings consist of triangles, squares, hexagons, octagons and dodecagons.* See Figure 1.7.3 for the eight classes of such tilings.

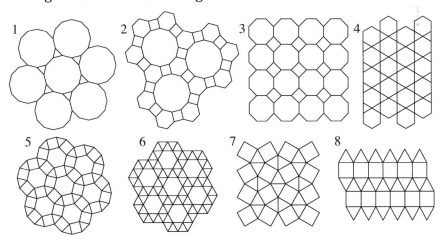

Figure 1.7.3

Proof. We first show that the number of polygons that may share a vertex in a semiregular tiling is 3, 4, 5, or 6. Suppose k polygons share a common vertex. Clearly $k \geq 3$. Let $\theta_1, \theta_2, \ldots, \theta_k$ denote the vertex angles of the polygons. Since each θ_i is at least $\pi/3$ we have $2\pi = \theta_1 + \theta_2 + \cdots + \theta_k \geq k\pi/3$, so that $k \leq 6$.

Since semiregular tilings are uniform, at each vertex we have k regular polygons with $n_1, n_2, \ldots, n_k$ sides, $3 \leq k \leq 6$, and each $n_i \geq 3$. Since the vertex angles of the polygons measure $(n_i - 2)\pi/n_i$ we have $\sum_{i=1}^{k} (n_i - 2)\pi/n_i = 2\pi$, or equivalently

$$\sum_{i=1}^{k} \frac{1}{n_i} = \frac{k-2}{2}.$$

The solutions to these four equations ($k = 3, 4, 5, 6$) are given in Table 1.1.

TABLE 1.1

k	n_1	n_2	n_3	n_4	n_5	n_6	Conclusion
3	3	7	42				✗
	3	8	24				✗
	3	9	18				✗
	3	10	15				✗
	3	12	12				semiregular (1)
	4	5	20				✗
	4	6	12				semiregular (2)
	4	8	8				semiregular (3)
	5	5	10				✗
	6	6	6				regular
4	3	3	4	12			not uniform
	3	3	6	6			semiregular (4)
	3	4	4	6			semiregular (5)
	4	4	4	4			regular
5	3	3	3	3	6		semiregular (6)
	3	3	3	4	4		semiregular (7, 8)
6	3	3	3	3	3	3	regular

The six solutions for $k = 3$ marked ✗ do not yield tilings, since in a tiling if one of $\{n_1, n_2, n_3\}$ is odd, then the other two must be equal, as those tiles alternate around the tile with an odd number of sides. Also, the solution (3,3,3,4,4) corresponds to two semiregular tilings, one with the two squares adjacent and one where they are not adjacent.

One of the solutions for $k = 4$ is indicated as "not uniform." In Challenge 1.4 you can show that this is indeed the case. ∎

In the following chapters we relax the restrictions in the definitions of regular and semiregular tilings, and as a result we obtain a variety of tilings with congruent irregular convex polygons. Many of the tilings are neither uniform nor edge-to-edge.

Monohedral tilings in the Alhambra

In Figure 1.7.4 we see drawings of portions of two monohedral tilings in the Alhambra palace in Granada, Spain. The tiles in the one on the left are concave octagons and the tiles in the one on the right are concave dodecagons.

Figure 1.7.4

1.8. Voronoi diagrams and dual tilings

A ***Voronoi diagram***, named for the Ukrainian mathematician Georgy Feodosevich Voronoi (1868–1908), is a partition of the plane consisting of a set S of points p and regions $\text{Vor}(p)$ consisting of points in the plane closer to p than to any other point in S. Voronoi-like diagrams also appeared in the work of René Descartes (1596–1650) and Peter Gustav Lejeune Dirichlet (1805-1859). Today Voronoi diagrams find application in a variety of fields, such as pattern recognition, crystallography, facility location, cartography, etc. See Figure 1.8.1 for a photograph of Voronoi and an example of a Voronoi diagram for a set S with 11 points.

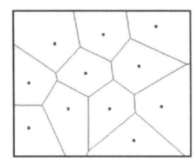

Figure 1.8.1

Note that each region is a convex polygon whose sides are portions of the perpendicular bisectors of the line segments joining pairs of points in S. See [Devadoss and O'Rourke, 2011] for further information.

Of interest to us is the case where the set S is the set of vertices of a regular or semiregular tiling of the plane. When S consists of the vertices of one of the three regular tilings in Figure 1.7.2, we have the three Voronoi diagrams in Figure 1.8.2 (in black), where the dots are the vertices of the regular tiling with gray edges.

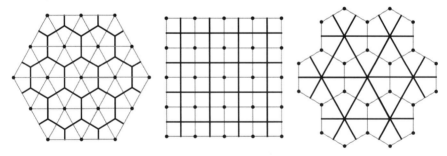

Figure 1.8.2

The Voronoi diagram for a tiling of the plane is another tiling, called the corresponding **_dual tiling_** (the vertices of the tiling in the Voronoi diagram are the centers of the polygons in the regular tiling). Observe that the regular tilings with triangles and hexagons are duals of one another, while the tiling with squares is self-dual.

Conspicuously missing from the tilings in the previous section are tilings with pentagons. While regular pentagons do not appear in regular or semiregular tilings, a variety of irregular ones do appear in monohedral tilings with irregular pentagons, as we show in the next chapter. When we look at the Voronoi diagram for a set S consisting of the vertices of semiregular tiling where five tiles meet at each vertex (e.g., tilings 6 and 7 in Figure 1.7.3), we obtain the Voronoi diagrams in Figure 1.8.3.

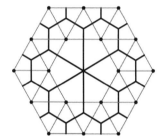 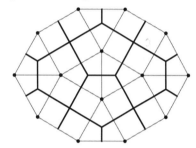

Figure 1.8.3

The resulting pentagonal tiling on the left is called a *floret tiling*, with pentagons having four 120° angles and one 60° angle. The one on the right is called a *Cairo tiling*, whose pentagons have two 90° angles and three 120° angles. See Section 2.6 for more about monohedral pentagonal tilings.

A third pentagonal tiling results from the Voronoi diagram for semiregular tiling 8 in Figure 1.7.3. See Challenge 1.5.

1.9. Challenges

Each chapter in this book concludes with *Challenges* for the reader. The challenges provide an opportunity for you to use and build upon the mathematics of polygons presented in the chapter. Solutions to all the challenges in the book can be found after Chapter 8.

1. POLYGON BASICS

1.1 Let p_n and P_n (k_n and K_n) denote respectively the perimeters (areas) of regular n-gons inscribed in and circumscribed about the same circle, as in Theorem 1.4.1. Show that

$$p_n/P_n = \cos\frac{\pi}{n} \text{ and } k_n/K_n = \cos^2\frac{\pi}{n}.$$

1.2 Let a_n and A_n denote the side lengths of the inscribed and circumscribed n-gons in Theorem 1.4.1. Show that

$$A_{2n} = \frac{a_n A_n}{a_n + A_n} \text{ and } a_{2n} = \sqrt{a_n A_{2n}/2}.$$

1.3 Let a_n and A_n denote the side lengths of regular n-gons inscribed in and circumscribed about a circle with radius ρ. Show that

$$a_{2n} = \sqrt{2\rho^2 - \rho\sqrt{4\rho^2 - a_n^2}} \text{ and } A_{2n} = \frac{2\rho A_n}{2\rho + \sqrt{4\rho^2 + A_n^2}}.$$

1.4 Show that the solution (3,3,4,12) in Table 1.1 produces two tilings with vertices where two triangles, a square, and a dodecagon meet, but in each case there are also vertices of a second type. (Hint. Consider two cases, one where the two triangles are adjacent at a vertex, and one where they are not adjacent.)

1.5 Show that the Voronoi diagram for a set of vertices of the type 8 semiregular tiling in Figure 1.7.3 yields a monohedral tiling with convex pentagons.

1.6 In Section 1.2 we defined the interior angles of a polygon. The exterior angles of a convex polygon are angles formed by one side and the extension of an adjacent side as illustrated in Figure 1.10.1. (We won't consider exterior angles for a concave polygon, as the extension of a side adjacent to a reflex angle lies inside the polygon.)

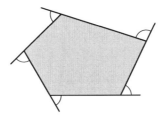

Figure 1.10.1

Show that the exterior angles of a convex polygon sum to 360°.

1.7 Show that among the interior angles of a convex polygon there cannot be more that three acute angles, and if a convex polygon has more than six sides, then at least one side has obtuse angles at both ends. (Hint. Consider the exterior angles.)

1.8 Let $AB \cdots N$ be a regular n-gon with side length 1 and consider the $n-2$ triangles in $\triangle ABN$ determined by BN and the diagonals at A, as shown in gray in Figure 1.10.2. Show that for each of these triangles, the length of one of the sides is equal to the product of the lengths of the other two sides. (Hint. Consider a pair of adjacent triangles in $\triangle ABN$.)

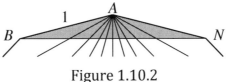

Figure 1.10.2

1.9 Figure 1.6.1 shows that within each regular star polygon $\{n/k\}$ there is a small regular n-gon similar to the circumscribing regular n-gon with side length a. Show that for the stars $\{2n/(n-1)\}$ (e.g., $\{6/2\}$, $\{8/3\}$, $\{10/4\}$, etc.) the side length of the inner n-gon is $a\tan(\pi/2n)$.

CHAPTER 2

Pentagons

A Pentagon ... having casually discovered the constituents of the simpler colours and a rudimentary method of painting, is said to have begun decorating first his house, then ... lastly himself. The convenience as well as the beauty of the results commended themselves to all.

Edwin Abbott Abbott
Flatland: A Romance of Many Dimensions

2.1. Introduction

Pentagons (or 5-*gons*) are five-sided polygons, and exist in a variety of shapes. In Figure 2.1.1 we see eleven examples [Grünbaum, 1975]. The first four in the top row are *simple* pentagons—the first one is *convex* (and *regular*). We devote the next three sections to properties of regular pentagons, and Section 2.5 to irregular convex pentagons. The next three pentagons in the first row are *concave*. The concave pentagons differ in the number (one or two) and location (adjacent or nonadjacent) of reflex angles (interior angles measuring more than 180°).

FIGURE 2.1.1

The remaining seven are examples of *complex* (or *crossed*) pentagons, the final one being the *pentagram*, the familiar star polygon {5/2} in Section 1.6. We discuss pentagrams in Section 2.7. Not pictured here are *skew* pentagons, those whose vertices are not all in the same plane. Note that all pentagons are *unicursal*, as compound pentagons do not exist.

Pentagons in nature

Approximations to regular pentagons can be found in nature. In Figure 2.1.2 we see a petunia blossom, a crystal of iron pyrite, also known as fool's gold, and a cross-section of okra.

FIGURE 2.1.2

2.2. Regular pentagons

Regular pentagons—convex pentagons with five sides of equal length and five equal vertex angles—have been studied since the time of the ancient Greek geometers. Their interest in them may be due in part to the connection between regular pentagons and the *golden ratio* $\varphi = (1 + \sqrt{5})/2$. From Section 1.3 we have the following data for regular pentagons (recall that r and R are the inradius and circumradius, respectively):

vertex angle	$\theta = 3\pi/5 = 108°$
side length	$a = 2r \tan 36° = 2R \sin 36°$
semiperimeter	$s = 5a/2$
area	$K = \dfrac{5}{4}a^2 \cot 36° = 5r^2 \tan 36°$
	$= \dfrac{5}{2}R^2 \sin 72° = rs$

2.2. REGULAR PENTAGONS

circumradius $\quad R = \frac{a}{2}\csc 36°$

inradius $\qquad r = \frac{a}{2}\cot 36° = R\cos 36°$

To make use of the above data, it will be advantageous to evaluate the trigonometric functions at angles with measure $36°$ ($\pi/5$) and $72°$ ($2\pi/5$). To do so we first examine the diagonals of a regular pentagon.

Theorem 2.2.1. *The length of each diagonal of a regular pentagon with side length 1 equals the golden ratio $\varphi = (1+\sqrt{5})/2$.*

Proof. Let x denote the length of each diagonal, as shown Figure 2.2.1a. Applying Ptolemy's Theorem 1.4.2 to the shaded quadrilateral yields $x^2 = x + 1$, and hence x equals the positive root of $x^2 - x - 1 = 0$, i.e., $x = \varphi$. Note that $\varphi^2 = \varphi + 1$. ∎

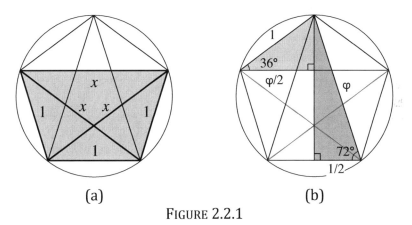

FIGURE 2.2.1

Example 2.2.1. *Pentagon trigonometry.* Since the diagonals trisect each vertex angle, the marked angles in Figure 2.2.1b have the indicated measures. Hence in the light gray right triangle we have $\cos 36° = \varphi/2 = (1+\sqrt{5})/4$, and in the dark gray right triangle we have $\cos 72° = 1/(2\varphi) = (\sqrt{5}-1)/4$.

Trigonometric identities now yield $\sin 36° = \sqrt{3-\varphi}/2 = \sqrt{10-2\sqrt{5}}/4$, $\tan 36° = \sqrt{7-4\varphi} = \sqrt{5-2\sqrt{5}}$, and $\sin 72° = \sqrt{2+\varphi}/2 = \sqrt{10+2\sqrt{5}}/4$. Hence we have the following data:

$$\text{side length} \quad a = R\sqrt{3-\varphi}$$

$$\text{area} \quad K = 5r^2\sqrt{7-4\varphi} = 5R^2\sqrt{2+\varphi}/4$$

$$\text{inradius} \quad r = R\varphi/2, \text{ etc. } \square$$

Example 2.2.2. *The golden ratio is irrational.* The first two numbers the ancient Greeks proved irrational were $\sqrt{2}$ and φ, the lengths of the diagonals in a square and a regular pentagon each with side 1. The proofs are indirect, showing that it is impossible for the number to be rational. Assume φ is rational, i.e., $\varphi = m/n$ where m and n are integers with $0 < n < m$, and the fraction is in lowest terms. Then there exists a regular pentagon with side n and diagonal m, as shown in Figure 2.2.2.

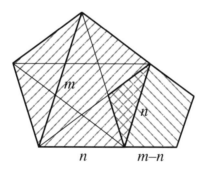

FIGURE 2.2.2

It now follows that there is a smaller regular pentagon with side $m-n$ and diagonal n, with $m-n < n$ (since $m < 2n$), $n < m$, and $\varphi = n/(m-n)$. This contradicts our assumption that the fraction m/n was in lowest terms; hence φ is irrational. $\square$

2.2. REGULAR PENTAGONS

The golden ratio is omnipresent in mathematics

The number φ we call the *golden ratio* or *golden section* was the *extreme and mean ratio* to the ancient Greeks and the *divine proportion* to 16th century Italian mathematician Luca Pacioli. The German mathematician Martin Ohm attached the adjective *golden* to *ratio* and *section* in 1835. Since then there has been much written about its occurrence in nature, art, and architecture. But what is certain is its occurrence in mathematics. We have seen φ in the geometry and trigonometry associated with the regular pentagon. It also appears in elementary algebra ($x^5 \pm 1 = (x \pm 1)(x^2 \mp \varphi x + 1)(x^2 \pm x/\varphi + 1)$), integral calculus ($\pi\varphi/5 = \int_0^\infty dx/(1+x^{10})$), hyperbolic functions (arcsinh$(1/2) = \ln\varphi$), complex analysis ($2\sin(i\ln\varphi) = i$), and number theory (the nth Fibonacci number is the integer nearest to $\varphi^n/\sqrt{5}$).

Example 2.2.3. *Golden triangles and gnomons.* A *golden triangle* is an isosceles triangle in which the ratio of one of the two equal sides to the third side is φ, i.e., it is similar to a triangle formed by one side and two diagonals of a regular pentagon. The angles of a golden triangle measure 36°, 72°, and 72°. Note that the pentagram in Figure 2.1.1 consists of a central pentagon surrounded by five golden triangles.

A *golden gnomon* is an isosceles triangle in which the ratio of one of the two equal sides to the third side is $1/\varphi$, i.e., it is similar to a triangle formed by two sides and a diagonal of a regular pentagon. The angles of a golden gnomon measure 36°, 36°, and 108°. See Challenge 2.6. □

In the next theorem we present a nice result relating the sums of distances to the five vertices of a regular pentagon from an arbitrary point on its circumcircle.

Theorem 2.2.2. *If ABCDE is a regular pentagon and P is any point on its circumcircle between A and B, then $PA + PB + PD = PC + PE$. See Figure 2.2.3.*

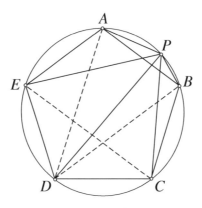

FIGURE 2.2.3

Proof. Let a be the side length and $a\varphi$ the length of each diagonal. Applying Ptolemy's theorem to quadrilateral $PBDA$ yields $a \cdot PD = a\varphi \cdot PA + a\varphi \cdot PB$ so that $PA + PB = PD/\varphi$. Applying Ptolemy's theorem to quadrilateral $PCDE$ yields $a\varphi \cdot PD = a \cdot PC + a \cdot PE$ so that $PC + PE = \varphi PD$. Hence

$$PA + PB + PD = \frac{PD}{\varphi} + PD = \frac{1+\varphi}{\varphi} PD = \varphi PD = PC + PE. \blacksquare$$

Fire hydrant pentagons

Next time you pass a fire hydrant, take a close look at the nuts on the top and sides. They are shaped like regular pentagons, as seen in Figure 2.2.4. The nuts have this shape, rather than being square or hexagonal, so that they cannot be loosened or tightened with an ordinary wrench. Firefighters use a special tool for the task.

FIGURE 2.2.4

2.3. Drawing a regular pentagon

Perhaps the earliest known method for drawing a regular pentagon is the one in Euclid's *Elements*, Proposition 11 in Book IV. It is not a simple construction, so we present several others that are. In our first example we have a construction from Ptolemy of Alexandria's *Almagest*, the one the second author encountered in his high school geometry course.

Example 2.3.1. *Ptolemy's construction*. Draw a circle with center O and diameter BC as shown in Figure 2.3.1. Draw a radius AO perpendicular to BC. Let D be the midpoint of BO and draw AD. Locate E in CO such that $DE = AD$. Then AE is the side length of a regular pentagon inscribed in the circle.

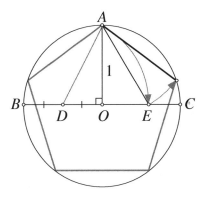

FIGURE 2.3.1

To show that the construction is valid, let the radius AO of the circle equal 1. Then $DO = 1/2$ and $AD = \sqrt{5}/2$. Hence $EO = (\sqrt{5}-1)/2 = \varphi - 1$ and $AE = \sqrt{3-\varphi}$, the correct side length (see Example 2.2.1) when $R = 1$. □

In the next example we have another construction, published by Herbert William Richmond [Richmond, 1873].

Example 2.3.2. *Richmond's construction*. Let AO and BO be perpendicular radii in a circle as shown in Figure 2.3.2, let C be the midpoint of AO, and draw BC. Bisect ∠BCO, let the bisector intersect BO at D, and draw DE perpendicular to BO. Then BE is a side of the inscribed regular pentagon.

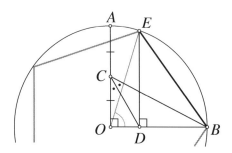

FIGURE 2.3.2

To show that the construction is valid, let $AO = BO = 1$, so that $CO = 1/2$ and $BC = \sqrt{5}/2$. By the angle bisector theorem, $BD/DO = BC/CO = \sqrt{5}$, and hence $DO = (\sqrt{5}-1)/4$. Thus $\angle BOE = \arccos(\sqrt{5}-1)/4 = 72°$, so that BE is a side of the inscribed regular pentagon. □

For our final example we present a construction from 1864 by Yosifusa Hirano [Fukagawa and Pedoe, 1989]. It is in the tradition of the *sangaku* ("mathematical tablet"), geometry theorems that were written on wooden tablets during the Edo period (1603-1867) in Japan and hung on Buddhist temples and Shinto shrines as offerings. Many sangaku featured tangent circles, as does Hirano's construction.

Example 2.3.3. *Hirano's construction.* Let AB and CD be perpendicular diameters of a circle with center O as shown in Figure 2.3.3. Draw circles with diameters AO and BO, let P be the midpoint of BO, and draw CP intersecting the circle with center P at T. Let the circle with center C and radius CT intersect the circle with center O at E and F. Then EF is the side of a regular pentagon inscribed in the circle with center O.

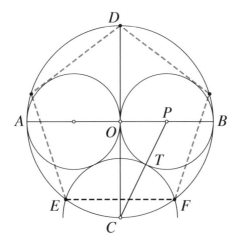

FIGURE 2.3.3

To show that the construction is valid, assume the radius of the circle with center O is 1. Then $CP = \sqrt{5}/2$ so that $CF = CT = (\sqrt{5}-1)/2$. Hence $\sin \angle CDF = CF/CD = (\sqrt{5}-1)/4$ so that $\angle CDF = 18°$, $\angle COF = 36°$, and $\angle EOF = 72°$. Note that the construction also yields EC and CF as sides of a regular decagon. □

A variety of other constructions are possible, see [Štěpánová, 2017].

Pentagonal logos

Pentagons are frequently used as corporate logos, highway signs, military insignia, etc. Regular pentagons are popular, as are pentagons with right angles. Here are a few examples.

FIGURE 2.3.4

2.4. Folding a regular pentagon

If a ribbon or a narrow strip of paper of constant width is tied into a simple overhand knot and carefully tightened and flattened, the knot looks like the one in Figure 2.4.1. The knot is clearly a pentagon, but is it *regular*? Most likely not for real world ribbon, but we answer for mathematical ribbon, which has zero thickness.

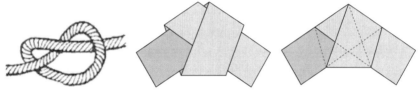

FIGURE 2.4.1

We examine what happens when the ribbon is folded. In Figure 2.4.2a we see that at each fold line, the angles of incidence and reflection are equal, and that the triangle with heavier lines as sides is isosceles at each fold.

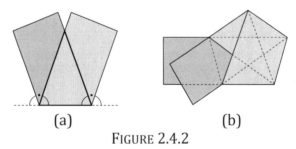

FIGURE 2.4.2

After folding the knot as in Figure 2.4.1 fold the end of the ribbon on the right in Figure 2.4.1 behind (or in front) of the knot, as shown in Figure 2.4.2b. The sides of the ribbon in the final fold will be parallel to the bottom edge by the equality of the angles of incidence and reflection, creating the final diagonal in the pentagon. Note that since the ribbon has constant width, each diagonal is parallel to a side of the pentagon. With folds on four sides of the pentagon all the interior angles are equal. All five of the isosceles triangles with two diagonals as sides (the golden triangles) are congruent, so the pentagon is equilateral, and hence regular.

Meccano and Erector models of regular pentagons

Meccano is a popular construction toy invented by Frank Hornby in 1898 in England, consisting of metal strips with evenly spaced holes, nuts and bolts, and other pieces. An *Erector* set is a similar toy invented by A. C. Gilbert in 1913 in the United States. In addition to making models of bridges, buildings, vehicles, etc., these sets can make models of regular polygons, as demonstrated by Gerard 't Hooft, a physics Noble laureate, in his article *Meccano Math* ['t Hooft, 2009].

Below we see a 1922 advertisement for Erector sets, and a model of a regular pentagon (in dark gray) made from six strips of length 2, two of length 3 and one each of lengths 4, 5, and 6. See 't Hooft's article for other constructions.

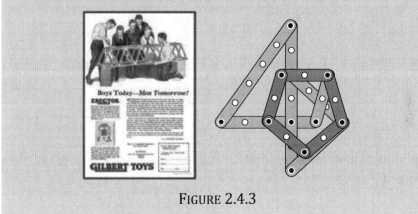

FIGURE 2.4.3

2.5. Six types of general convex pentagons

We begin with some vocabulary for types of pentagons with specified properties. *Cyclic*, *tangential*, *equilateral*, and *equiangular* pentagons are those with a circumcircle, an incircle, sides of equal length, and equal angles, respectively. Two other types refer to properties of the diagonals. An *equidiagonal* pentagon is one whose diagonals have equal length. A *parallel* pentagon is one in which each diagonal is parallel to a side.

Clearly regular pentagons are cyclic, tangential, equilateral, equiangular, equidiagonal, and parallel. Figure 2.5.1 illustrates cyclic, tangential, and equiangular irregular pentagons. For the cyclic irregular pentagon, chose any five points on a circle except ones evenly spaced, and the same for a tangential irregular pentagon. Drawing line segments parallel to the sides of a regular pentagon produces an equiangular irregular pentagon.

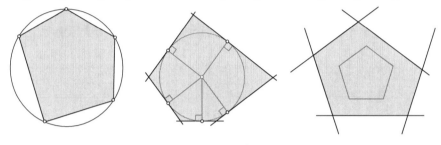

FIGURE 2.5.1

In the next three examples we illustrate constructions for equilateral, equidiagonal and parallel irregular pentagons.

Example 2.5.1. *Construction of equilateral irregular pentagons.* To construct an equilateral irregular pentagon $ABCDE$ with sides a, draw $AB = a$ as seen in Figure 2.5.2a. Chose C on a circle of radius a and center B, and chose E on a circle of radius a and center A with the length of the line segment CE (not shown) less than $2a$. Then let D be the intersection of circular arcs with radius a from C and D. □

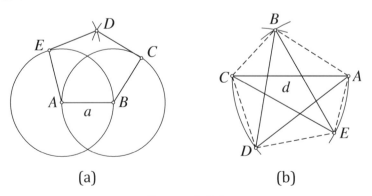

(a)   (b)

FIGURE 2.5.2

2.5. SIX TYPES OF GENERAL CONVEX PENTAGONS

Example 2.5.2. *Construction of equidiagonal irregular pentagons.* This construction is analogous to the one in the preceding example. To construct an equidiagonal irregular pentagon *ABCDE* with diagonals of length *d*, draw $AC = d$ as shown in Figure 2.5.2b. Chose *D* on a circular arc with radius *d* and center *A*, and chose *E* on a circular arc with radius *d* and center *C*. Then let *B* be the intersection of circular arcs with radius *d* from *D* and *E*. The sides of *ABCDE* are the dashed lines in the figure. □

Example 2.5.3. *Construction of irregular parallel pentagons.* Figure 2.5.3 shows that if we stretch a regular pentagon we produce an irregular parallel pentagon (here we have stretched the pentagon horizontally by 50%).

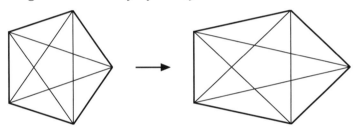

FIGURE 2.5.3

For a construction using properties of the golden ratio, observe in Figure 2.5.4 that the pentagon with vertices (0,0), (1,0), (φ, 1), (1, φ), and (0,1) is an irregular parallel pentagon. □

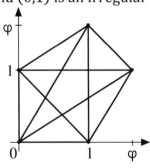

FIGURE 2.5.4

Satisfying two (or more) of the six properties may force a pentagon to be regular. An equilateral equiangular pentagon is regular, by definition. An equilateral cyclic pentagon is also regular, as is an equiangular tangential pentagon. See Theorems 7.5.1 and 7.6.1.

The next theorem presents a nice area property of parallel pentagons.

Theorem 2.5.1. *The diagonals of a pentagon cut off triangles with equal area if and only if it is a parallel pentagon.*

Proof. See Figure 2.5.5. Triangles *ABC* and *ABE* have equal area if and only if they have the same altitudes, i.e., if and only if *CE* is parallel to *AB*. Triangles *ABC* and *BCD* have the same area if and only if they have the same altitudes, i.e., if and only if *AD* is parallel to *BC*. Similarly for the remaining pairs of triangles with a common base, hence all five triangles have the same area if and only if *ABCDE* is a parallel pentagon. ∎

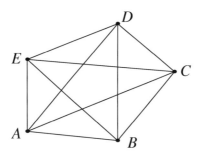

FIGURE 2.5.5

Baseball's pentagon

The description of baseball's home plate from Major League Baseball's website mlb.com reads as follows: *Home plate is a* 17-*inch square of whitened rubber with two of the corners removed so that one edge is* 17 *inches long, two adjacent sides are* 8-1/2 *inches each and the remaining two sides are* 12 *inches each and set at an angle to make a point.* See Figure 2.5.6 for the pentagon so described.

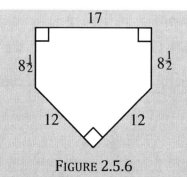

FIGURE 2.5.6

But something is wrong here! The description implies that there is an isosceles right triangle with legs 12 inches and hypotenuse 17 inches. And there isn't, since $12^2 + 12^2 = 288 \neq 289 = 17^2$.

2.6. Pentagonal tilings

In Section 1.7 we showed that regular pentagons do not appear in either regular or semiregular tilings of the plane. However, they do form the surface of a regular dodecahedron (see Section 2.8). In addition a variety of irregular convex pentagons do yield monohedral pentagonal tilings. The story of the discovery of all 15 such tilings is an interesting one [Schattschneider, 1975, 1981; Gardner, 1988 (Ch. 13)], we give only a brief summary here.

The story begins with the 1918 doctoral dissertation of Karl Reinhardt [Reinhardt, 1918], in which he identified five monohedral pentagonal tilings and three hexagonal ones. The following description of Reinhardt's pentagonal tilings is from [Kershner, 1968]. Denote the angles of the hexagon by A, B, C, D, and E, (in that order); and the sides by $a = EA$, $b = AB$, $c = BC$, $d = CD$, and $e = DE$. Then a convex pentagon $ABCDE$ tiles the plane if it belongs to one or more of the following classes:

Type 1. $A + B + C = 360°$;

Type 2. $A + B + D = 360°$ and $a = d$;

Type 3. $A = C = D = 120°$ and $a = b, d = c + e$;

Type 4. $A = C = 90°$ and $a = b, c = d$;

Type 5. $A = 60°, C = 120°, a = b, c = d$.

The next figures illustrate each type of tiling pentagon and a portion of the tiling pattern.

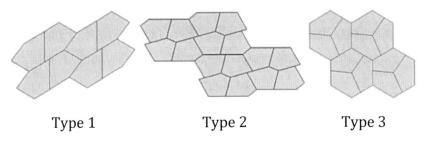

Type 1   Type 2   Type 3

FIGURE 2.6.1

Note that any convex pentagon with a pair of parallel sides is a Type 1 tiling pentagon.

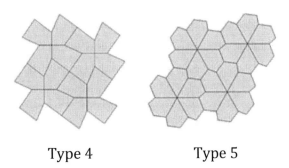

Type 4   Type 5

FIGURE 2.6.2

Example 2.6.1. *An equilateral irregular pentagonal tiling.* If we use equal side lengths in a Type 4 tiling, we obtain the attractive pattern in Figure 2.6.3a. This tiling is known as the *Cairo tiling* for its reported use in that city. Many other equilateral pentagons tile the plane; for details see [Schattschneider, 1978; Hirschhorn and Hunt, 1985]. □

2.6. PENTAGONAL TILINGS

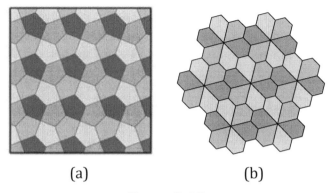

(a) (b)

FIGURE 2.6.3

Example 2.6.2. *The floret pentagonal tiling.* Setting $B = D = E = 120°$ and $c = d = e$ in a Type 5 tiling yields the tiling illustrated in Figure 2.6.3b. Its name follows from the observation that clusters of six tiles resemble flower petals. □

In [Kerschner, 1968] the author included three more classes of tiling pentagons, illustrated in Figure 2.6.4.

Type 6. $A + B + D = 360°, A = 2C$, and $a = b = e, c = d$;

Type 7. $2B + C = 2D + A = 360°$ and $a = b = c = d$;

Type 8. $2A + B = 2D + C = 360°$ and $a = b = c = d$.

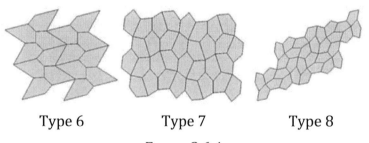

Type 6   Type 7   Type 8

FIGURE 2.6.4

In a 1975 article in the "Mathematical Games" column in the *Scientific American*, Martin Gardner presented the eight pentagonal tilings above, noting Kershner's opinion that the list of pentagon types that tile the plane was complete. Intrigued by this, readers (many of whom were not mathematicians) endeavored to find others. The first two to do so were Marjorie Rice, a homemaker and amateur mathematician (Type 9), and Richard James, a computer scientist (Type 10) in 1975-76.

Type 9. $2A + C = D + 2E = 360°, b = c = d = e$;

Type 10. $A = 90°, B + 2C = 360°, B + E = 180°, a = b = c = e$.

 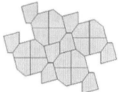

Type 9   Type 10

FIGURE 2.6.5

Example 2.6.3. *The origin of Type 10.* Richard James discovered Type 10 by starting with the semiregular tiling with octagons and squares, deleting the squares, shifting the octagons, partitioning the octagons into four pentagons, and then inserting copies of the pentagons into the gap between the rows of octagons, as shown in Figure 2.6.6 [Schattschneider, 1981]. □

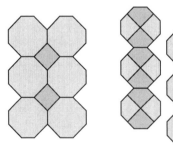 

FIGURE 2.6.6

In 1976-77 Marjorie Rice found three more types, as illustrated in Figure 2.6.7, bringing the total number of classes of tiling pentagons to thirteen.

Type 11. $A = 90°, 2B + C = 360°, C + E = 180°, 2a + c = d = e$;

Type 12. $A = 90°, 2B + C = 360°, C + E = 180°, 2a = d = c + e$;

Type 13. $B + E + 90°, 2A + D = 360°, d = 2a = 2e$.

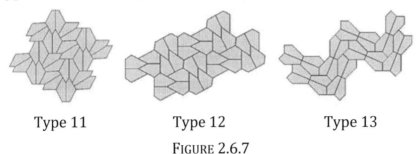

Type 11   Type 12   Type 13

FIGURE 2.6.7

The fourteenth convex pentagon tiling was discovered in 1985 by Rolf Stein of the University of Dortmund. In Figure 2.6.8 we see a portion of the tiling along with the cover of the November 1985 issue of *Mathematics Magazine*, which announced its discovery. Unlike the previous 13 types, the shape of this tile is completely determined by the angle and side specifications.

Type 14. $A = 90°, 2B + C = 360°, C + E = 180°,$
           $2a = 2c = d = e.$

Type 15. $A = 150°, B = 60°, C = 135°, D = 105°, E = 90°,$
           $a = c = e, b = 2a.$

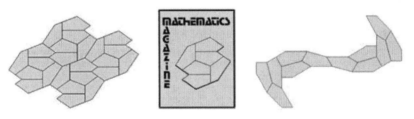

Type 14   Type 15

FIGURE 2.6.8

In 2015 a fifteenth type was discovered by mathematicians Casey Mann, Jennifer McLoud-Mann, and David Van Derau at the University of Washington-Bothell. In Figure 2.6.8 we see a portion of the tiling that generates the entire tiling using only translations.

See [Rao, 2017] for a computer-assisted proof that there are no other types of convex pentagons that tile the plane.

2.7. Pentagrams

The rightmost pentagon in the second row of pentagons in Figure 2.1.1 is a *regular pentagram* or *pentacle*, the regular star polygon with five sides with symbol {5/2}. The followers of Pythagoras in the 6th century BCE used the pentagram as their symbol. Its side length d (the length of a line segment joining two adjacent vertices of the pentagram) equals $a\varphi$, as it is a diagonal of the regular convex pentagon with the same vertices and side length a. Here are some data for the pentagram with vertex angle θ and side d. Note that the circumradius R is the same for both the convex regular pentagon and its inscribed pentagram, but that the inradius r for the two are different.

vertex angle	$\theta = \pi/5 = 36°$
side length	$d = a\varphi = 2R \sin 72° = R\sqrt{2+\varphi}$
circumradius	$R = \dfrac{d}{2} \csc 72° = r \sec 72° = 2\varphi r$
inradius	$r = \dfrac{d}{2} \cot 72° = R \cos 72° = \dfrac{R}{2\varphi}$

Thus the sum of the five vertex angles of a regular pentagram is 180°. However, that is also true for irregular pentagrams, as the following figure illustrates [Nahkli, 1986].

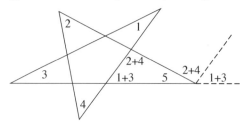

FIGURE 2.7.1

In the next theorem we present several expressions of the area K of a regular pentagram in terms of the golden ratio and one of the linear measures associated with the pentagram. The expressions involve the *Lucas numbers* L_n defined by $L_1 = 1$, $L_2 = 3$, and $L_n = L_{n-1} + L_{n-2}$ for $n \geq 3$. Recall that the area of pentagram is the area of the gray star in Figure 2.7.2.

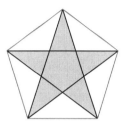

FIGURE 2.7.2

Theorem 2.7.1. *If $K = [\{5/2\}, a]$ denotes the area of a regular pentagram $\{5/2\}$ with side length d, circumradius R, and inradius r inscribed in a regular pentagon with side length a, then*

$$K = \frac{1}{2}a^2\sqrt{5(7-4\varphi)} = \frac{5}{2}R^2\sqrt{18-11\varphi}$$
$$= \frac{1}{2}d^2\sqrt{5(47-29\varphi)} = 10r^2\sqrt{3-\varphi}.$$

Proof. From (1.8) the area $K = [\{5/2\}, a]$ of the pentagram $\{5/2\}$ is $(5/2)a^2 \cot 72°$. Since $\cot 72° = (1/5)\sqrt{5(7-4\varphi)}$ we have $K = (1/2)a^2\sqrt{5(7-4\varphi)}$, so that $K \cong 0.8123a^2$. Using $a = R\sqrt{3-\varphi}$ and some algebra yields $K = 5R^2\sqrt{18-11\varphi}/2$, so that $K \cong 1.12257R^2$. Using $R^2 = d^2/(2+\varphi)$ and more algebra yields $K = d^2\sqrt{5(47-29\varphi)}/2$, so that $K \cong 0.31027d^2$. Lastly, since $R = 2\varphi r$ we have $K = 10r^2\sqrt{3-\varphi}$ so that $K \cong 11.75571r^2$.

The appearance of the first eight Lucas numbers 1, 3, 4, 7, 11, 18, 29, and 47 in the exact expressions for K is not surprising, since the nth Lucas number L_n is approximately φ^n. ∎

Corollary 2.7.2. *The ratio of the area of a regular pentagram to the area of the regular pentagon with the same vertices is $4\varphi - 6 = 2\sqrt{5} - 4 \cong 0.472136$.*

Proof. The areas of the pentagram and pentagon are $(5/2)R^2\sqrt{18-11\varphi}$ and $(5/4)R^2\sqrt{2+\varphi}$, respectively, hence the desired ratio is

$$2\sqrt{\frac{18-11\varphi}{2+\varphi}} = \frac{2}{\sqrt{5}}\sqrt{(18-11\varphi)(3-\varphi)} = \frac{2}{\sqrt{5}}(7-4\varphi) = 4\varphi - 6. \blacksquare$$

Example 2.7.1. *How the pentagram partitions the pentagon.* Figure 2.7.2 shows that the diagonals of a regular pentagon not only form a regular pentagram, but also partition the pentagon into 11 regions: five golden triangles (the points of the star), five golden gnomons (the white triangles), and one smaller pentagon. The techniques from the proof of Theorem 2.7.1 show that if the area of the original pentagon is 1, then the area of each point of the star is $(7\varphi - 11)/5 \cong 0.06525$, the area of each white gnomon is $(7 - 4\varphi)/5 \cong 0.10557$, and the area of the small pentagon is $5 - 3\varphi \cong 0.14590$. $\square$

Pentagrams in religious art and architecture

Many world religions have employed the pentagram as a symbol. On the left in Figure 2.7.3 we see a Roman stone carving from the Temple of Jupiter in Split, Croatia. In the center is a well in the Seimei Shrine in Kyoto, Japan. This Shinto shrine was founded in 1007 in memory of Abe no Seimei.

FIGURE 2.7.3

On the right we see a pentagram in the exterior of a rose window in the north transept of the Cathédrale Notre-Dame d'Amiens in Amiens, France, built in the 13th century.

2.7. PENTAGRAMS

Example 2.7.2. *Lengths of line segments in general pentagrams.* As with vertex angles, there is a nice relationship for lengths of line segments in a general pentagram. In Figure 2.7.4a we see pentagram $ABCDE$ in which we have labeled the sides of the small triangle with vertex A as a_1, a_2, and a_3 and similarly for the other vertices. With the segments so labeled we have

$$a_1 b_1 c_1 d_1 e_1 = a_2 b_2 c_2 d_2 e_2.$$

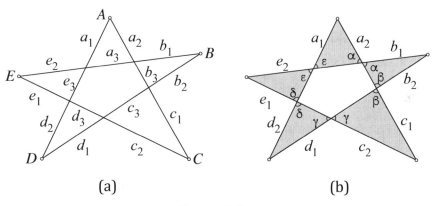

(a)   (b)

FIGURE 2.7.4

For a proof we apply the sine law to each of the gray triangles in Figure 2.7.3b yielding $a_1/a_2 = \sin\alpha/\sin\varepsilon$, $b_1/b_2 = \sin\beta/\sin\alpha$, $c_1/c_2 = \sin\gamma/\sin\beta$, $d_1/d_2 = \sin\delta/\sin\gamma$, and $e_1/e_2 = \sin\varepsilon/\sin\delta$. Multiplying the five equations yields $a_1 b_1 c_1 d_1 e_1/a_2 b_2 c_2 d_2 e_2 = 1$ [Lee, 1998]. □

Pentagrams on flags

The flags of many nations feature five-pointed stars, i.e., concave decagons. The list includes Brazil, Cameroon, Chile, China, Cuba, Myanmar, Panama, Senegal, Somalia, Turkey, New Zealand, the United States, Venezuela, Vietnam, and many others as well as the European Union. However, currently (2022) only two countries have a national flag with a regular pentagram— Ethiopia and Morocco—as seen in Figure 2.7.5.

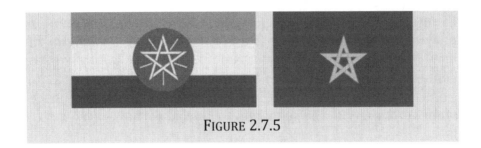

FIGURE 2.7.5

2.8. Pentagons in space

As mentioned in Section 2.6, regular pentagons form the surface of a regular *dodecahedron*, a convex polyhedron with 12 identical regular pentagonal faces, as illustrated in Figure 2.8.1a. Another solid with pentagonal faces is the *hexagonal truncated trapezohedron* (with 12 regular pentagonal and 2 regular hexagonal faces) in Figure 2.8.1b.

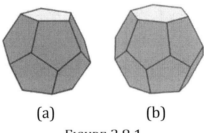

(a)   (b)

FIGURE 2.8.1

In Figure 2.8.2 we see a *truncated icosahedron*, a convex polyhedron with 12 regular pentagonal and 20 regular hexagonal faces. It is a model for the soccer ball, and for the carbon molecule C_{60} known as *buckminsterfullerene*, discovered in 1985 and named for the American architect Richard Buckminster Fuller (1895-1983) whose geodesic domes the molecule's structure resembles.

FIGURE 2.8.2

2.8. PENTAGONS IN SPACE

What do the three polyhedra have in common? In each, faces are pentagons or hexagons, and three faces meet at each vertex. In the following theorem we show that if a convex polyhedron satisfies those two properties, it must have exactly 12 pentagonal faces. In the theorem the polygons need not be regular.

Theorem 2.8.1. *If every face of a convex polyhedron is either a pentagon or a hexagon and three faces meet at each vertex, then the polyhedron has exactly twelve pentagonal faces.*

Proof. Let V, E, and F denote the total number of vertices, edges, and faces, respectively, and F_5 and F_6 the number of pentagonal and hexagonal faces, respectively. Then $V - E + F = 2$ by Euler's polyhedral formula, $F_5 + F_6 = F$ and $5F_5 + 6F_6 = 2E$. Since three edges meet at each vertex and each edge joins two vertices, we have $2E = 3V$. But $6V - 6E + 6F = 12$, hence $6F - 2E = 12$. Thus $6(F_5 + F_6) - (5F_5 + 6F_6) = 12$ and so $F_5 = 12$. ∎

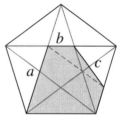

FIGURE 2.8.3

Example 2.8.1. *Dürer's solid.* On the left in Figure 2.8.3 we see a copy of the 1514 engraving *Melencolia I* by Albrecht Dürer (1471–1528). The polyhedron in the engraving is often called *Dürer's solid.*

While it is difficult to be sure, it may well be a *truncated rhombohedron*, a convex polyhedron with two equilateral triangular faces and six irregular pentagonal faces. The pentagons are truncated rhombi found within a regular pentagon. Its angles measure 72°, 108°, 126°, 126°, 108°, and sides $a = 1, b = 2 - \varphi, c = \sqrt{7 - 4\varphi}$. □

A great many other polyhedra have regular and irregular pentagons (along with other polygons) as faces.

2.9. Miscellaneous examples

In this section we present several additional examples employing pentagons.

Example 2.9.1. *The Sphinx concave pentagon*. The *sphinx* is a name given to a concave pentagon composed of six equilateral triangles for its vague resemblance to the ancient structure in Egypt. See Figure 2.8.1a.

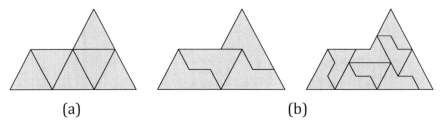

FIGURE 2.9.1

The sphinx is a *reptile* (for *rep*licating *tile*)—a polygon that can be dissected into congruent copies of itself. In Figure 2.9.1b we see the sphinx dissected into four and nine copies of itself. The sphinx also tiles the plane in a variety of ways, two of which are illustrated in Figure 2.9.2. □

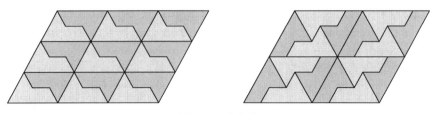

FIGURE 2.9.2

2.9. MISCELLANEOUS EXAMPLES

Example 2.9.2. *A dodecahedron model.* Drawing the diagonals in a regular pentagon produces a smaller central regular pentagon, as illustrated in Figure 2.9.3a. Erasing those diagonals and drawing diagonals in the smaller pentagon and extending them to the edges of the original large pentagon yields five more small pentagons each congruent to the one in the center, as seen in Figure 2.9.3b.

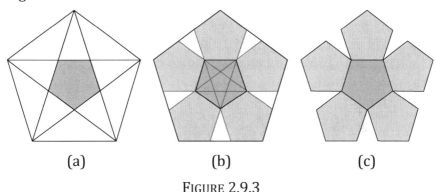

(a)   (b)   (c)

FIGURE 2.9.3

If we draw two copies of the pentagons in Figure 2.9.3c on cardboard and cut them out, we have the surface of a regular dodecahedron. Score each along the edges of the central pentagon, place one copy of six pentagons behind the other, and weave a rubber band over and under the ten corners, as shown in Figure 2.9.4a.

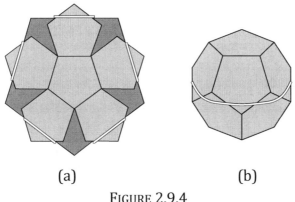

(a)   (b)

FIGURE 2.9.4

If the rubber band is sufficiently tight, the model will pop up to form an approximate regular dodecahedron, as seen in Figure

2.9.4b. The design for the pentagons in this model was employed by the United States Postal Service in 2000 for a set of five pentagonal $1 stamps with the theme "Exploring the Solar System." □

2.10. Pentagons in architecture

Perhaps the world's best-known—and largest—pentagonal building is *The Pentagon*, home of the U.S. Department of Defense in Arlington County, Virginia, as seen in the photograph in Figure 2.10.1. Built during World War II, it has 3.7 million square feet of office space, 17.5 miles of corridors, and employs 26,000 people.

FIGURE 2.10.1

But it is not the only such structure. Notable others are the 16th century *Villa Farnese* in the province of Viterbo, Italy, in Figure 2.10.2a; and the 18th century *Mole Vanvitelliana* in Ancona, Italy, in Figure 2.10.2b.

(a)　　　　　　　　　　　　(b)

FIGURE 2.10.2

2.10. PENTAGONS IN ARCHITECTURE

A pentagonal home in *Flatland*

At the beginning this chapter we had a quote from Edwin Abbott's 1884 novel *Flatland, A Romance of Many Dimensions*, in which not only were the characters polygons, but they lived in polygonal homes. In Figure 2.10.3 we see Abbott's drawing of his narrator's pentagonal house.

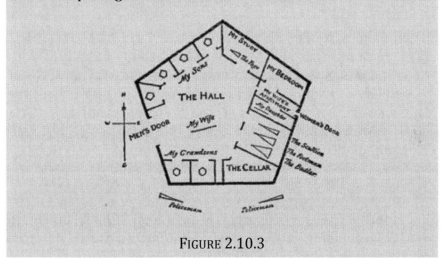

FIGURE 2.10.3

2.11. Challenges

2.1 Let *ABCDE* be a regular pentagon inscribed in a unit circle. Show that the product of the lengths of two sides and two diagonals from a vertex is 5, i.e.,

$$AB \cdot AC \cdot AD \cdot AE = 5.$$

2.2 A *clipped rectangle* is a convex pentagon formed by cutting off one corner of an $a \times b$ rectangle to form a fifth side of length *c*, as shown in Figure 2.11.1. If *d* and *e* are the lengths of the two diagonals nearest *c*, show that

$$a^2 + b^2 + c^2 = d^2 + e^2.$$

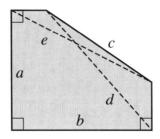

FIGURE 2.11.1

2.3 *A Putnam pentagon.* A convex pentagon $P = ABCDE$, with vertices labeled consecutively, is inscribed in a circle of radius 1. Find the maximum area of P subject to the condition that the chords AC and BD be perpendicular. (This is problem A-4 on the 1984 William Lowell Putnam Mathematical Competition.)

2.4 Show that any polygon can be tiled by convex pentagons.

2.5 Suppose that for the parallel pentagon $ABCDE$ in Figure 2.5.5, each of the five triangles cut off by a diagonal has area 1. Show that the area of $ABCDE$ is $2 + \varphi$.

2.6 Figure 2.2.1a shows that a regular pentagon can be dissected into a variety of golden triangles and golden gnomons. Use that fact and the golden gnomon in Figure 2.11.2 to derive the following infinite series:

$$1 + \frac{1}{\varphi^2} + \frac{1}{\varphi^4} + \cdots = \varphi.$$

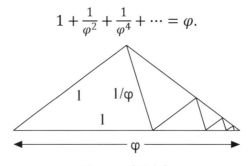

FIGURE 2.11.2

2.7 Show that (i) a cyclic equilateral pentagon is regular, and (ii) a cyclic equiangular pentagon is regular.

2.11. CHALLENGES

2.8 Let *ABCDE* be a regular pentagon, and let *F* be the intersection of *BD* and *CE*, as shown in Figure 2.11.3. Show that the concave pentagon *BCDEF* can tile the plane.

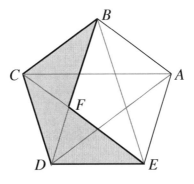

FIGURE 2.11.3

2.9 Among the many puzzles created by Sam Loyd (1841-1911) one finds the "Smart Alec" puzzle [Loyd, 1914]. The polygon in Figure 2.11.4a is a concave pentagon formed by removing an isosceles right triangle from a square. The task proposed by Loyd was to dissect the pentagon into four pieces that could be reassembled to form a square. Loyd's solution is shown in Figure 2.11.4b

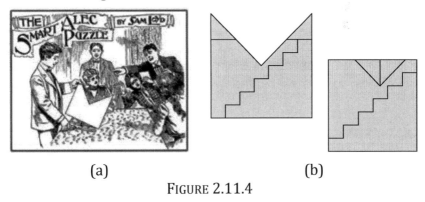

(a) (b)

FIGURE 2.11.4

(i) Show that the "square" in Loyd's solution is actually a non-square rectangle.

(ii) No four-piece solution is known. Show that it can be done with five pieces.

2.10 *Loyd's pentagram puzzle.* Another Loyd puzzle is "The New Star" in which Loyd asks us "to show how and where to place another star of the first magnitude" in Figure 2.11.5. By "first magnitude" Loyd means larger than all the others. Solve the puzzle.

FIGURE 2.11.5

CHAPTER 3

Hexagons

Bees...by virtue of a certain geometrical forethought...know that the hexagon is greater than the square and the triangle and will hold more honey for the same expenditure of material.

Pappus of Alexandria
Mathematical Collection, Book V

She could not fathom the hexagonal miracle of snowflakes formed from clouds, crystallized fern and feather that tumble down to light on a coat sleeve, white stars melting even as they strike. How did such force and beauty come to be in something so small and fleeting and unknowable?

Eowyn Ivey
The Snow Child

3.1. Introduction

Hexagons are six-sided polygons (or *6-gons*) and, like the pentagons in the preceding chapter, come in a variety of shapes. Below are a few examples of convex, concave, complex, and skew hexagons and common names for some of them. Figure 3.1.1 illustrates the convex case.

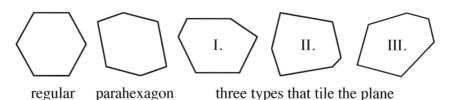

regular   parahexagon   three types that tile the plane

FIGURE 3.1.1

Regular hexagons are equilateral and equiangular, and are the subject of the next section. *Parahexagons* have three pairs of parallel sides of equal length, and are the subject of Section 3.5. We discuss convex hexagons that tile the plane in Section 3.4.

In Figure 3.1.2 we see two examples of equilateral concave hexagons, a concave parahexagon, and an example of an "L" hexagon, with five right angles. L-*polyominoes* are interesting L-hexagons and are discussed in Section 3.7.

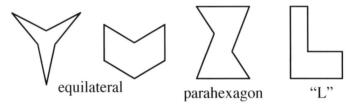

FIGURE 3.1.2

The complex, or crossed hexagons have a variety of shapes. In Figure 3.1.3 we have placed dots at the six vertices to distinguish them from the places where edges cross. In Figure 3.1.3 the vertices coincide with the vertices of a regular hexagon, but that is not necessary. There is one compound hexagon—the union of two triangles—the regular case illustrated is the *hexagram*. See Section 3.8.

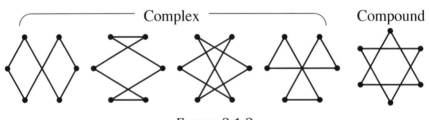

FIGURE 3.1.3

In Figure 3.1.4 we see two examples of skew hexagons using six edges of a cube and of a regular octahedron. Each one is "regular" in the sense that it is both equilateral and equiangular, but we shall reserve the term "regular hexagon" for the planar ones.

3.1. INTRODUCTION

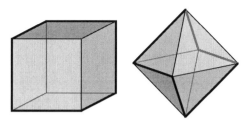

FIGURE 3.1.4

The skew hexagons in Figure 3.1.4 are the *Petrie polygons* of two Platonic solids, the cube and the octahedron. They consist of edges of the solid such that every two (but no three) consecutive edges belong to the same face. They are named for the British mathematician John Flinders Petrie (1907–1972).

Regular hexagons in nature

The quotes at the beginning of this chapter about honeycombs and snowflakes remind us that these are among the best-known occurrences of approximate regular hexagons in nature. In the figure below we see photograph of a honeybee and honeycomb, and a macro photograph of a real snowflake.

FIGURE 3.1.5

3.2. Regular hexagons

The regular hexagon is perhaps the regular polygon that is the easiest to draw with a straightedge and compass. Draw a circle and a diameter, and using the radius of the circle draw two arcs using the diameter endpoints as centers. Then the six points so obtained on the circle are the vertices of a regular hexagon. See Figure 3.2.1.

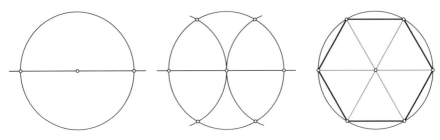

FIGURE 3.2.1

Drawing two more diameters partitions the hexagon into six congruent equilateral triangles, which enables us to easily derive the following data for a regular hexagon with side length a, circumradius R, and inradius r.

$$
\begin{aligned}
\text{vertex angle} \quad & \theta = 2\pi/3 = 120° \\
\text{side length} \quad & a = R = 2r\sqrt{3}/3 \\
\text{semiperimeter} \quad & s = 3a = 3R \\
\text{area} \quad & K = 3a^2\sqrt{3}/2 = 2r^2\sqrt{3} \\
& = 3R^2\sqrt{3}/2 = rs = 3ar \\
\text{circumradius} \quad & R = a \\
\text{inradius} \quad & r = a\sqrt{3}/2 = R\sqrt{3}/2
\end{aligned}
$$

Another way to draw a regular hexagon is to trisect the sides of an equilateral triangle and connect the middle thirds of each side, as shown in Figure 3.2.2a.

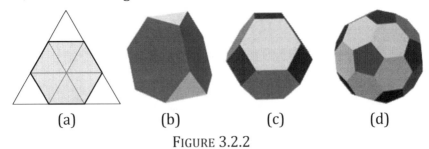

(a) (b) (c) (d)

FIGURE 3.2.2

3.2. REGULAR HEXAGONS

No Platonic solid has hexagonal faces, but three of them—the regular tetrahedron, octahedron, and icosahedron—have equilateral triangles as faces. Truncating each of those solids truncates the triangular faces, yielding three Archimedean solids with regular hexagonal faces, as illustrated in Figures 3.2.2b (the *truncated tetrahedron*, with four regular hexagonal and four equilateral triangular faces), 3.2.2c (the *truncated octahedron*, with eight regular hexagonal and six square faces), and 3.2.2d (the *truncated icosahedron*, with 20 regular hexagonal and 12 regular pentagonal faces). A great many other polyhedra also have regular hexagonal faces.

Hexagonal logos and signs

As with pentagons, hexagons are often used for corporate logos, military insignia, and traffic signs. Here are a few examples.

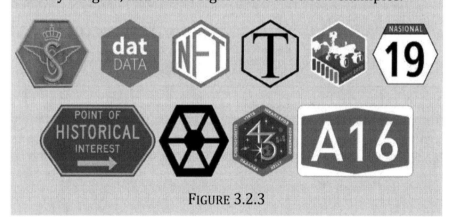

FIGURE 3.2.3

The regular hexagon with side length a has nine diagonals as shown in Figure 3.2.4a. The long three are diameters of length $2a$ (solid lines) and the short six, with length $a\sqrt{3}$ (dashed lines), are the sum of the lengths of the altitudes of a pair of adjacent triangles. Note that the length of a short diagonal equals twice the inradius r.

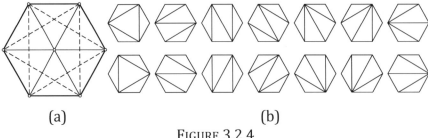

(a)   (b)

FIGURE 3.2.4

Choosing three diagonals that do not intersect in the interior of the hexagon "triangulates" the hexagon, partitioning it into four triangles. There are exactly 14 ways to do that, as illustrated in Figure 3.2.4b. Leonhard Euler (1707-1783) posed the problem of counting the number of ways to triangulate a convex n-gon in a letter to Christian Goldbach (1690-1764) in 1751. The answer is C_{n-2} for $n \geq 3$, where $C_n = \frac{1}{n+1}\binom{2n}{n}$ is the nth *Catalan number* (sequence A000108 in the *Online Encyclopedia of Integer Sequences* at oeis.org) named for Eugène Charles Catalan (1814-1894). For a proof see [Honsberger, 1973].

In the next theorem we present a nice result relating the sums of distances to the six vertices of a regular hexagon from an arbitrary point on its circumcircle. It is a companion to Theorem 2.2.2 for a regular pentagon.

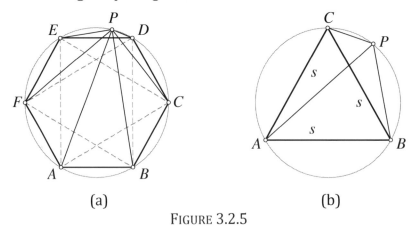

(a)   (b)

FIGURE 3.2.5

3.2. REGULAR HEXAGONS

Theorem 3.2.1. *If ABCDEF is a regular hexagon and P is any point on its circumcircle between D and E, then $PA + PB = PC + PD + PE + PF$.* See Figure 3.2.5a.

Before proving the theorem we prove a simple lemma, named for the Dutch mathematician Frans van Schooten (1615-1660), about distances to the vertices of an equilateral triangle from any point on its circumcircle.

Lemma 3.2.2 (Van Schooten's theorem). *If $\triangle ABC$ is an equilateral triangle and P is any point on its circumcircle, then the longest of the three segments PA, PB, and PC equals the sum of the other two.* See Figure 3.2.5b.

Proof. Let s denote the side length of $\triangle ABC$. Applying Ptolemy's theorem 1.4.2 to the quadrilateral $PBAC$ yields $s \cdot PA = s \cdot PB + s \cdot PC$, hence $PA = PB + PC$. ∎

Proof of Theorem 3.2.1. In Figure 3.2.5a applying Lemma 3.2.2 to P and $\triangle ACE$ yields $PA = PC + PE$ and to P and $\triangle BDF$ yields $PB = PD + PF$, from which the result follows. ∎

See Challenge 3.3 for another application of Lemma 3.2.2. The next two examples exhibit clever uses of regular hexagons.

Example 3.2.1. *Assigning numbers to points in the plane.* Problem 6 on the 2001 USAMO (United States of America Mathematical Olympiad) reads as follows.

> Each point in the plane is assigned a real number such that for any triangle, the number at the center of its inscribed circle is equal to the arithmetic mean of the three numbers at its vertices. Prove that all points in the plane are assigned the same number.

The following solution is by Michael Hamburg, St. Joseph's High School, South Bend, Indiana, who created an ingenious construction (see Figure 3.2.6) and then concisely and elegantly derived the result. For this he was awarded the Clay Olympiad Scholar Award.

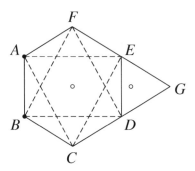

FIGURE 3.2.6

Let A and B be arbitrary points. Construct the regular hexagon $ABCDEF$ and let G be the intersection of CD and EF extended. Note that the figure is symmetric about the line through G parallel to AE and BD. Triangles ACE and BDF share the same incenter, as do CEG and FDG. Denote the number assigned to a point by its lower case letter. Then $a + c + e = b + d + f$ and $c + e + g = d + f + g$, hence $a = b$. □

Example 3.2.2. *Counting calissons.* Calissons are French sweets in the shape of a rhombus formed by two equilateral triangles. See Figure 3.2.7a. Calissons could come in a regular hexagonal box (but apparently do not) as shown in Figure 3.2.7b, packed with the short diagonal of the calisson parallel to one of the sides of the box, so that there are three possible orientations for each calisson.

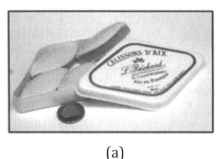 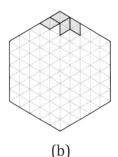

(a) (b)

FIGURE 3.2.7

3.2. REGULAR HEXAGONS

The next theorem tells us how many calissons in the box are in each of the three orientations when the box is full (i.e., tiled by calissons).

Theorem 3.2.3. *In any packing of calissons in a regular hexagonal box, the number of calissons in each of the three orientations is the same, and equal to one-third of the number of calissons in the box.*

Proof [David and Tomei, 1989]. In Figure 3.2.8a we see an arbitrary filling of the box with identical calissons, and in Figure 3.2.8b we have shaded the calissons in three different shades of gray. Now they appear as cubes in a room with a square floor and square sides. Viewing the configuration from above we see only the top faces of the cubes, which of course cover the floor. The same is true if we view the configuration from one of the sides. Hence the number of cube faces—i.e., calissons—is the same in each of the three orientations. ∎

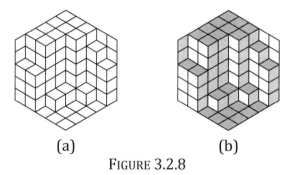

FIGURE 3.2.8

Example 3.2.4. *Tying a hexagonal knot.* In Section 2.4 we saw how to form a pentagon by tying an overhand knot in a strip of paper. Tying a square knot using two strips of paper forms a knot that approximates a regular hexagon, as shown in Figure 3.2.9.

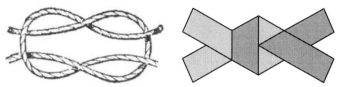

FIGURE 3.2.9

For more on polygonal knots, see [Morley, 1924; Brunton, 1961] □

Example 3.2.5. *The benzene molecule.* One of the most beautiful structures in organic chemistry is the benzene molecule C_6H_6, containing six carbon atoms and six hydrogen atoms arranged in a ring as shown in Figure 3.2.10.

FIGURE 3.2.10

It is often pictured with alternating single and double bonds between adjacent carbon atoms. Such bonds have different lengths, but actually all of the carbon-carbon bonds have the same length, so that the molecule is flat and shaped like a regular hexagon. □

Regular hexagons in arts and crafts

Regular hexagons have appeared in various art forms for centuries. In Figure 3.2.11 we see several examples. On the left is detail from a Roman mosaic in the Bardo National Museum in Tunis. In the center is a ceramic tile from the late 15th century, currently in the Metropolitan Museum of Art in New York City. On the right is a 19th century American quilt, also in the Metropolitan Museum of Art.

FIGURE 3.2.11

3.3. Cyclic hexagons

Many irregular hexagons are also cyclic, and in this section we present some of their properties.

Theorem 3.3.1. *If ABCDEF is a cyclic hexagon, then $A + C + E = B + D + F = 2\pi$.*

Proof. See Figure 3.3.1, noting that the angles A, C, and E together subtend arcs (*subtend* means "to be opposite to") totaling twice the circumference of the circumcircle, as do B, D, and F. Hence $A + C + E = B + D + F = 2\pi$ (since the sum of all six vertex angles is 4π). ∎

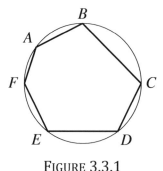

FIGURE 3.3.1

Example 3.3.1. *Area and circumradius of a cyclic hexagon.* If the side lengths of a cyclic hexagon are x, x, x, y, y, y ($x \neq y$) in that order, find (i) the area K of the hexagon and (ii) its circumradius R. See Figure 3.3.2a.

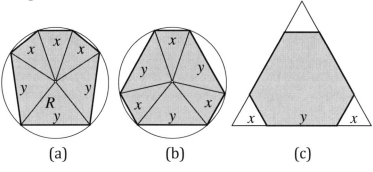

FIGURE 3.3.2

Rearrange the order of the sides to obtain the hexagon in Figure 3.3.2b. Note that each vertex angle is now $2\pi/3$ with the area and circumradius of the hexagon unaltered. The new hexagon is a truncated equilateral triangle as shown in Figure 3.3.2c. Hence the area of the original hexagon is

$$K = [(2x+y)^2 - 3x^2]\frac{\sqrt{3}}{4} = (x^2 + 4xy + y^2)\frac{\sqrt{3}}{4}$$

(recall that the area of an equilateral triangle with side length a is $\sqrt{3}a^2/4$). See Challenge 3.5 for the circumradius R. □

Theorem 3.3.2. *If ABCDEF is a cyclic hexagon, then the long diagonals are concurrent if and only if*

$$AB \cdot CD \cdot EF = BC \cdot DE \cdot FA.$$

Proof. Consider a cyclic hexagon $ABCDEF$ as shown in Figure 3.3.3. First note that $\triangle AOB \sim \triangle EOD$ (since the marked angles subtend the same arc) and thus

$$\frac{AB}{DE} = \frac{p}{u+y} = \frac{q}{s+z}.$$

Similarly $\triangle CPD \sim \triangle APF$ and $\triangle EQF \sim \triangle CQB$ so that

$$\frac{CD}{FA} = \frac{r}{p+z} = \frac{s}{v+x} \quad \text{and} \quad \frac{EF}{BC} = \frac{u}{r+x} = \frac{v}{q+y}.$$

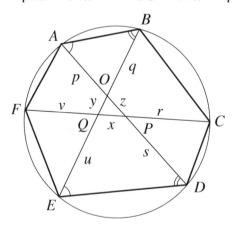

FIGURE 3.3.3

3.3. CYCLIC HEXAGONS

Hence

$$\left(\frac{AB \cdot CD \cdot EF}{BC \cdot DE \cdot FA}\right)^2 = \frac{p}{p+z} \cdot \frac{q}{q+y} \cdot \frac{r}{r+x} \cdot \frac{s}{s+z} \cdot \frac{u}{u+y} \cdot \frac{v}{v+x},$$

and the right side equals 1 if and only if $x = y = z = 0$. ∎

In 1890 the German mathematician Wilhelm Ferdinand Fuhrmann (1833-1904) published an extension of Ptolemy's theorem to cyclic hexagons, relating the six side lengths and the lengths of the three long diagonals.

Fuhrmann's Theorem 3.3.3. *Let ABCDEF be a cyclic quadrilateral with side lengths a, a', b, b', c, c' and long diagonals with lengths e, f, g; so chosen that a, a', e have no common vertex, nor do b, b', f and c, c', g (see Figure 3.3.4a). Then*

$$efg = aa'e + bb'f + cc'g + abc + a'b'c'.$$

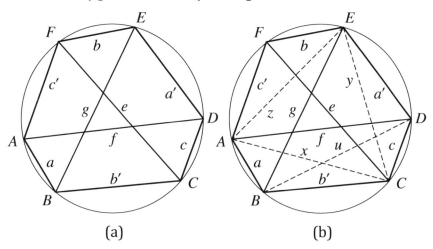

FIGURE 3.3.4

Proof [Johnson, 1960]. Draw the four short diagonals with lengths u, x, y, z as shown in Figure 3.3.4b. Applying Ptolemy's theorem to the cyclic quadrilaterals $ABDE$, $ACEF$, $BCDE$, and $ABCD$ yields, respectively, $fg = aa' + uz$, $ez = c'y + bx$, $uy = a'b' + cg$, and $ux = b'f + ac$. Hence

$$efg = aa'e + uze = aa'e + c'uy + bxu$$
$$= aa'e + a'b'c' + cc'g + bb'f + abc. \blacksquare$$

Example 3.3.2. *The circumradius of a cyclic parahexagon.* If ABCDEF in Theorem 3.3.3 is a cyclic parahexagon (see the definition of parahexagon in the paragraph following Figure 3.1.1), then $a' = a$, $b' = b$, $c' = c$, and $e = f = g = 2R$, as illustrated in Figure 3.3.5. Hence by Fuhrmann's theorem 3.3.3 the circumdiameter $d = 2R$ satisfies the cubic $d^3 - (a^2 + b^2 + c^2)d - 2abc = 0$. Also note that the long diagonals are concurrent by Theorem 3.3.2.

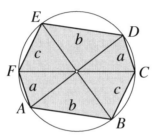

FIGURE 3.3.5

To illustrate, suppose $a = 1$, $b = \sqrt{5}$, and $c = \sqrt{2}$. Then $d^3 - 8d - 2\sqrt{10} = 0$, which factors as

$$(1/4)(d - \sqrt{10})(2d + \sqrt{10} - \sqrt{2})(2d + \sqrt{10} + \sqrt{2}) = 0$$

(very few cubics factor this easily). Hence $d = \sqrt{10}$ and $R = \sqrt{10}/2$. $\square$

There exist cyclic hexagons all of whose sides and diagonals are integers. See Challenge 3.12.

The game of Hex

Hex is a board game of strategy played on a grid of congruent regular hexagons in the shape of a rhombus. In Figure 3.3.6 we see an 11 by 11 version. It was invented by the Danish mathematician Piet Hein (1905-1996) in 1942 and rediscovered by the American mathematician John Nash (1928-2015) in 1948. Each player has one of two colors (e.g., black or white) and alternate placing a disk of their color on the board. The goal is to form a path from one border of their color to the other, the first player to do so wins.

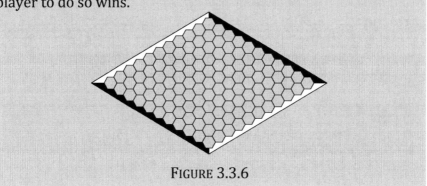

FIGURE 3.3.6

3.4. Hexagonal tilings

In the preceding chapter we saw that although regular pentagons do not tile the plane, a variety of irregular convex pentagons do. The situation with hexagons is different: Regular hexagons do tile the plane (as noted in the hexagonal grid of the Hex board in the preceding section), and just three types of convex irregular hexagons also tile the plane.

In Figure 3.4.1 we see a photograph of the beautiful regular hexagonal tiles designed by the Catalan architect Antoni Gaudí (1852-1926) that can be seen on the sidewalks along the Passeig de Gràcia in Barcelona.

FIGURE 3.4.1

In his doctoral dissertation of 1918, Karl Reinhardt showed that any convex hexagon that tiles the plane belongs to one of three classes. The following description of those classes is from [Kershner, 1968]. Denote the angles of the hexagon by A, B, C, D, E, and F (in that order); and the sides by $a = FA, b = AB, \ldots$, $f = EF$. Then a convex hexagon $ABCDEF$ tiles the plane if and only if it belongs to one or more of the following classes:

I. $A + B + C = 360°$ and $a = d$;

II. $A + B + D = 360°$ and $a = d, c = e$;

III. $A = C = E = 120°$ and $a = b, c = d, e = f$.

Note that regular hexagons belong to all three classes. Figure 3.4.2 illustrates each type of tiling pattern.

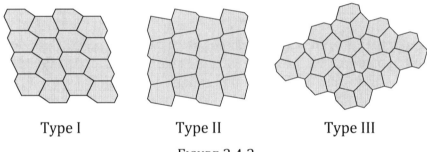

Type I   Type II   Type III

FIGURE 3.4.2

Semiregular tilings are obtained by combining two or more different types of regular polygonal tiles, as shown in Section 1.7. There are eight such tilings, four of which use hexagons. See Figure 3.4.3.

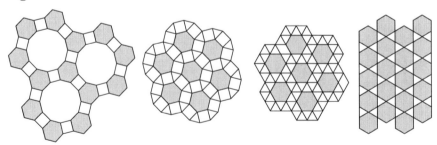

FIGURE 3.4.3

3.4. HEXAGONAL TILINGS

Example 3.4.1. *Hexagonal sections of cubes*. The 2011 Prova Cangur ("Kangaroo Examination") of the Societat Catalana de Matemàtiques included the following problem: A 3×3×3 cube consists of 27 small cubes. A plane passes through the center of the large cube perpendicular to a space diagonal of the large cube. How many small cubes intersect this plane?

The correct answer is 19. First note the regular hexagonal section of the large cube in Figure 3.4.4a, and that 8 of the 27 small cubes are not cut by the plane (4 can be seen in Figure 3.4.4b, 4 more are behind the hexagonal section).

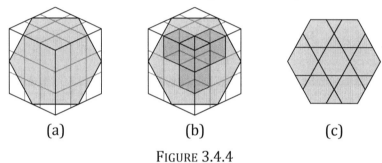

(a)　　　　　(b)　　　　　(c)

FIGURE 3.4.4

Alternatively one can count the regions of the hexagonal section in Figure 3.4.4c formed by the cubes intersected by the plane. Note that the tiling pattern on the hexagonal section is the rightmost semiregular tiling in Figure 3.4.3. □

3.5. Parahexagons

A *parahexagon* is a simple hexagon whose opposite sides are parallel and have equal length. Parahexagons may be convex or concave, as shown in Figures 3.1.1 and 3.1.2. Both types tile the plane; see Figure 3.5.1.

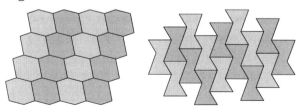

FIGURE 3.5.1

Example 3.5.1. *Centroid parahexagons*. In an arbitrary convex hexagon *ABCDEF*, connect the centroids of the six triangles *ABC*, *BCD*, *CDE*, *DEF*, *EFA*, and *FAB*, as shown in Figure 3.5.2a. The result, shaded gray, is a parahexagon.

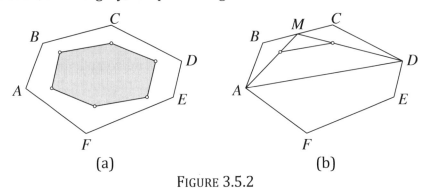

(a) (b)
FIGURE 3.5.2

Figure 3.5.2b shows why this is true [De Villiers, 2007]. Let *M* denote the midpoint of *BC*. Then the centroid of △*ABC* lies on *AM* 2/3 of the distance from *A* to *M*. Similarly the centroid of △*BCD* lies on *DM* 2/3 of the distance from *D* to *M*. Hence the line segment joining the two centroids is parallel to and 1/3 the length of the diagonal *AD*. Similarly the line segment joining the centroids of △*DEF* and △*EFA* with be parallel to and 1/3 the length of *AD*. Consequently the centroid hexagon is a parahexagon. □

Example 3.5.2. *The area of a median triangle*. Six copies of a triangle can be arranged to form a parahexagon, which enables us to use a parahexagon to derive triangle results. For example, the *median triangle* associated with an arbitrary triangle is one constructed from the three medians, as shown in Figure 3.5.3a.

We claim the area *M* of the median triangle is 3/4 of the area *T* of the original triangle. Six copies of the original triangle form the parahexagon of Figure 3.5.3b. The sides of the dashed triangle in Figure 3.5.3b have side lengths twice those of the three medians, hence its area is 4*M*. Since a median in each copy of the original triangle partitions it into two shaded triangles each with area $T/2$, we have $4M = 6(T/2) = 3T$. □

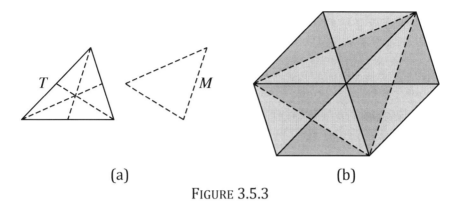

(a)   (b)

FIGURE 3.5.3

In Challenge 3.7 you can use a parahexagon to derive inequalities for the perimeters of a given triangle and its median triangle.

Example 3.5.3. *Another construction via triangles.* For a given triangle, construct three equilateral triangles using the sides of the given triangle, as shown in Figure 3.5.4a. Then the three equilateral triangles and three copies of the given triangle form a parahexagon, as seen in Figure 3.5.4b and its use as a tile in Figure 3.5.4c. □

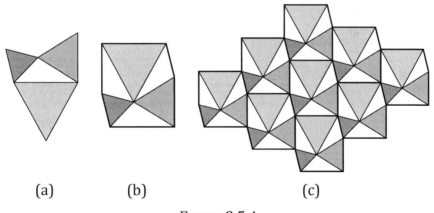

(a)   (b)   (c)

FIGURE 3.5.4

Next we use a parahexagon to prove a theorem about circles discovered by the American geometer Roger A. Johnson (1890-1954) in 1916 [Johnson, 1916]. It has been described as one of the few recent "really pretty theorems at the most elementary level of geometry" [Honsberger, 1976].

Johnson's Theorem 3.5.1. *If three circles with equal radii are drawn through a point, then the other three points of intersection determine a fourth circle with the same radius.* See Figure 3.5.5a.

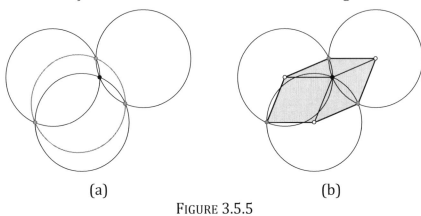

(a) (b)

FIGURE 3.5.5

Proof. Drawing radii in each circle to the intersection points creates three equilateral rhombi, which together form the gray equilateral parahexagon in Figure 3.5.5b.

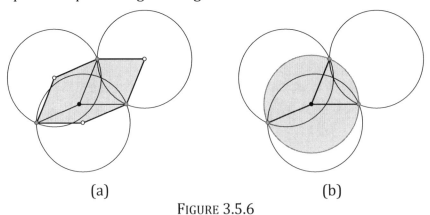

(a) (b)

FIGURE 3.5.6

Partitioning the parahexagon into three equilateral rhombi a second way locates a point equidistant (and equal to the common radius of the three circles) from the three pairwise intersections of the circles, a shown in Figure 3.5.6a. Hence the circle through the three pairwise intersections in Figure 3.5.6b has the same radius as in the original three circles. ∎

Convex parahexagons (and parallelograms) belong to a class of polygons called *parallelogons*—polygons with opposite sides parallel and equal that tile the plane. All parallelogons have 180° rotational symmetry, but perhaps less obvious is the fact that any parallelogon has only four or six sides.

Theorem 3.5.2. *Parallelograms and convex parahexagons are parallelogons. There are no other types of parallelogons.*

For a proof, see [Alexandrov, 2005, pp. 351-353].

3.6. The carpenter's square

A *carpenter's square* (or *framer's square* or *steel square*) is a tool used to mark lines and right angles in various construction and home improvement projects. It has a long wide arm (the *blade*) and a short narrow arm (the *tongue*), which meet at a right angle. See Figure 3.6.1a. A version appears along with compasses on the emblem of the Freemasons, as shown in Figure 3.6.1b.

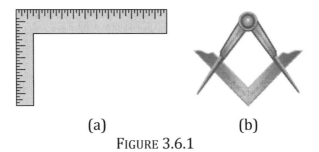

(a)　　　　　　　　(b)

FIGURE 3.6.1

The shape of the carpenter's square is a concave hexagon, the "L" hexagon in Figure 3.1.2. If we use the carpenter's square as a mathematical tool, like compasses, we can trisect angles [Richeson, 2017]. See Figure 3.6.2a.

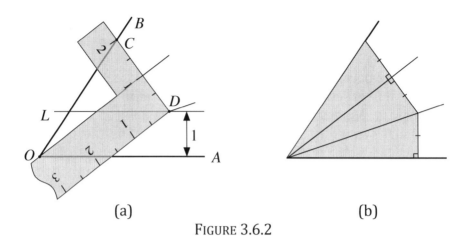

FIGURE 3.6.2

To trisect angle *AOB*, we use a carpenter's square whose blade is 1 unit wide, and draw line *L* parallel to and 1 unit above side *AO* of the triangle. Place the square on *AOB* with vertex *O* on the inside edge of the blade, the 2 unit mark of the tongue on *B* at point *C*, and the corner on *L* at *D*. Then the inside edge of the blade and the line *OD* trisect angle *AOB*, since the three triangles so formed, as shown in Figure 3.6.2b, are congruent.

In Example 4.5.3 we use a carpenter's square along with compass and straightedge to construct a regular heptagon, and in Example 6.2.1 we use a carpenter's square to construct a regular nonagon (a 9-gon) by trisecting the angles of an equilateral triangle inscribed in a circle.

3.7. L-polyominoes

Polyominoes are, like dominoes, polygons created by joining identical squares edge-to-edge. They have been studied since antiquity, although the name was coined by S. W. Golomb in 1953 (see [Golomb, 1994]). We assume that polyominoes are like dominoes, rigid polygons whose shape doesn't change when they are translated, rotated, of flipped over. In figure 3.7.1 we see the domino, the two trominoes, the five tetrominoes, and the twelve pentominoes.

3.7. L-POLYOMINOES

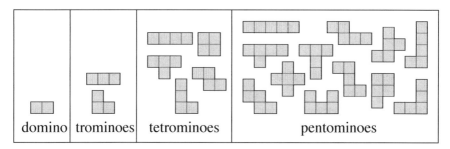

FIGURE 3.7.1

All polyominoes have an even number of edges. We are interested in the hexagonal polyominoes, all of which are concave L-hexagons, as in Figure 3.1.2. Some of them actually resemble the letter L, and are called L-*polyominoes*. To be precise, we have

Definition 3.7.1. *For any $n \geq 2$ an* L-polyomino *is a concave hexagon consisting of a column of n squares with another square attached at the bottom.*

Not all hexagonal polyominoes are L-polyominoes. For example, the two pentominoes in the lower right corner of Figure 3.7.1 are hexagonal but not L-pentominoes.

Many polyominoes tile the plane, but not all do. However, all the L-polyominos do tile, often in a variety of patterns. For example, Figure 3.7.2 illustrates three tilings with L-tetrominoes.

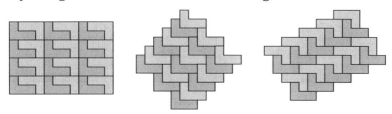

FIGURE 3.7.2

In addition to tiling the plane with polyominoes, mathematicians have considered tiling bounded regions in the plane, e.g., squares. A classic result in that vein is the following [Golomb, 1954].

Theorem 3.7.2. *If n is a power of two, then an n×n checkerboard with one square removed can be tiled by L-trominoes.*

Proof (by induction). For the base step, when *n* equals 2 the result is trivial. The induction step is illustrated in Figure 3.7.3. ∎

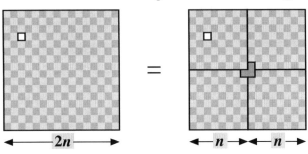

FIGURE 3.7.3

The common 8×8 checkerboard can be readily tiled with L-tetrominoes, as can many other squares and rectangles.

Theorem 3.7.3. *A rectangle with integer sides a and b, both greater than 1, can be tiled with L-tetrominoes if and only if $ab = 8m$ for some positive integer m.*

Proof [Golomb and Klarner, 1963]. Assume that the rectangle can be tiled with L-tetrominoes. Then 4 divides *ab*, so either *a* or *b* is even, suppose it is *a*. Consider a rectangle with *a* rows alternately colored black and white. Then every L-tetromino must cover three squares of one color and one square of the other color. If *m* L-tetrominoes cover 3 white squares and 1 black square and *n* L-tetrominoes cover 3 black squares and 1 white square, then $3m + n = ab/2 = 3n + m$, and hence $m = n$. Thus $4m = ab/2$, so $ab = 8m$. Conversely, if $ab = 8m$, then the rectangle can be dissected into smaller 2×4 and/or 3×8 rectangles each of which can be tiled with L-tetrominoes. ∎

Many more such tiling results can be found in [Golomb, 1994].

3.7. L-POLYOMINOES

In Example 2.9.1 we encountered the sphinx pentagon, a *reptile* or replicating tile, a polygon that can be dissected into copies of itself. Both the L-tromino and the L-tetromino are reptiles. In Figure 3.7.4 we see that the L-tromino (on the left) and the L-tetromino (on the right) can be dissected into 4, 9, and 16 copies of themselves. In fact, all L-polyominoes are reptiles; see Challenge 3.8.

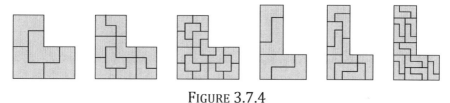

FIGURE 3.7.4

Example 3.7.1. *Using the L-tromino to sum an alternating series.* In Figure 3.7.5 we show that the partial sums of the series $1 - (1/2) + (1/4) - (1/8) + \cdots$ converge to $2/3$ using the fact that the L-tromino is a reptile.

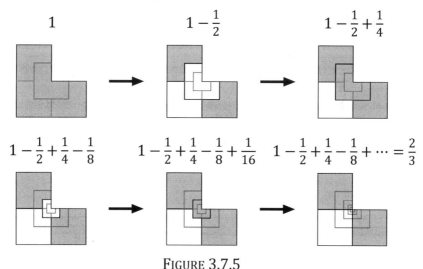

FIGURE 3.7.5

Example 3.7.2. *Perfect tilings, irreptiles and the golden bee.* In a reptile, all the small polygons are congruent and similar to the large one. In a *perfect tiling*, all the small polygons are similar to the large one, but no two are congruent. In this case the polygons are *irreptiles* (for *irregular reptiles*). The minimal case is when a perfect tiling uses just two irreptiles. There are only two such tilings [Schmerl, 2011], illustrated in Figure 3.7.6. One is triangular, the other hexagonal.

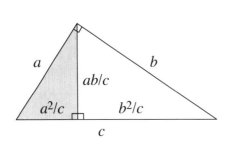 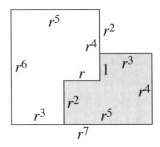

FIGURE 3.7.6

On the left we have a non-isosceles right triangle, tiled by two smaller similar but incongruent right triangles. On the right is one with similar concave hexagons, whose side lengths are in geometric progressions with common ratio r. Equating the lengths of the vertical sides of the gray hexagon yields $r^4 = r^2 + 1$, hence $r^2 = \varphi$, the golden ratio, and $r = \sqrt{\varphi} \cong 1.272$. Since the shape of the hexagon resembles the letter **b**, we have the name: *golden bee*. □

3.8. Hexagrams

The rightmost polygon in Figure 3.1.3 is a regular *hexagram*, the regular star polygon with six sides. It is a compound star, the union of two equilateral triangles. Its symbol is $\{6/2\} = 2\{3\}$. Some authors refer to compound star polygons such as the hexagram as *star figures* rather than star polygons. The regular hexagram is also called the *Star of David*, known worldwide as the symbol of the Jewish people. It also appears on the flag of Israel.

The edge length d of the regular hexagram is a diagonal of the regular hexagon with the same vertices and side length a, thus $d = a\sqrt{3}$. Here are some data for the hexagram with vertex angle θ and side length d.

3.8. HEXAGRAMS

vertex angle $\theta = \pi/3 = 60°$

side length $d = a\sqrt{3} = R\sqrt{3} = 2r\sqrt{3}$

area $K = ad = d^2/\sqrt{3} = a^2\sqrt{3} = R^2\sqrt{3}$

circumradius $R = a = d/\sqrt{3} = 2r$

inradius $r = R/2 = d/2\sqrt{3}$

The area of a regular hexagram is 2/3 the area of a regular hexagon with the same vertices, as shown in Figure 3.8.1a. Note that there are 18 congruent isosceles triangles in the hexagon of which 12 are in the hexagram. In Figure 3.8.1b we illustrate the area formula $K = ad$.

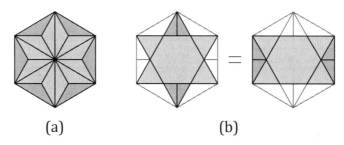

(a) (b)

FIGURE 3.8.1

Hexagrams in art

The hexagram has been used in art for centuries. In Figure 3.8.2 we have three examples. On the left we see a gold Frankish brooch from the late 8th or early 9th century, currently in the Cleveland Museum of Art.

FIGURE 3.8.2

In the center we have an image from the cover of the *Leningrad Codex*, an early 11th century handwritten Tanakh, or Hebrew bible, currently in the National Library of Russia in St. Petersburg. On the right we have a 17th century "Seal of Solomon" from the manuscript *Raphael, Oder Arzt-Engel* published in Amsterdam in 1676 by the German author Abraham von Franckenberg.

Example 3.8.1. *Magic hexagrams.* The reader may be familiar with magic squares—square arrays of integers with constant row and column "magic" sums. A *magic hexagram* is an arrangement of the integers 1 to 12 in the twelve triangular cells of the hexagram in Figure 3.8.3a so that the five numbers in each of the indicated six rows and diagonals have the same "magic" sum.

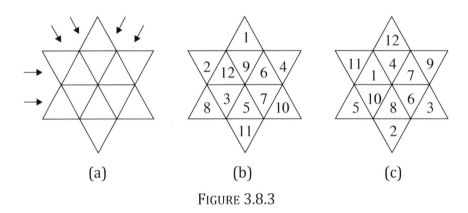

FIGURE 3.8.3

The solution, described in [Bolt et al., 1991; Gardner, 2000], concludes with a computer search that yields the solution in Figure 3.8.3b with magic sum 33, and its complement in Figure 3.8.3c with magic sum 32 (numbers in corresponding cells sum to 13). □

Example 3.8.2. *The unicursal hexagram and Pascal's theorem.* In Figure 3.8.4a we see a regular *unicursal hexagram*, whose sides are two long and four short diagonals of a regular hexagon. Although it is neither equilateral nor equiangular, we use the adjective regular since its vertices coincide with those of a regular hexagon.

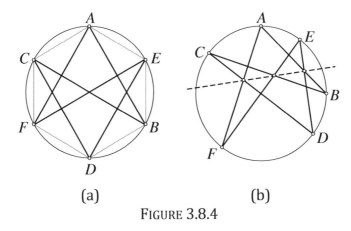

FIGURE 3.8.4

Every cyclic unicursal hexagram, regular or not, has an interesting property: the three points of intersection of pairs of opposite sides are collinear, as shown in Figure 3.8.4b (the pairs of opposite sides of a unicursal hexagon *ABCDEF* are *AB* and *DE*, *BC* and *EF*, and *CD* and *FA*).

At the age of sixteen the French mathematician Blaise Pascal (1623-1662) published a more general result in the following theorem, although his proof has never been found.

Pascal's Theorem 3.8.1. *If a hexagon is inscribed in a conic (e.g., a circle, ellipse, parabola, or hyperbola), then the three points of intersection of pairs of opposite sides (extended if necessary) lie on a straight line* (with suitable modifications if opposite sides are parallel).

Most proofs of the theorem employ projective geometry; see [van Yzeren, 1993; Augros, 2012]. The line through the three points of intersection is known as the *Pascal line* of the hexagon. For a cyclic irregular convex hexagon, the Pascal line lies outside the circumcircle, as shown in Figure 3.8.5. □

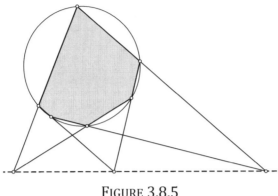

FIGURE 3.8.5

3.9. Miscellaneous examples

In this section we present a variety of results employing hexagons in the statement of a theorem or its proof.

Example 3.9.1. *The regular hexagon and lunes.* A *lune* (from the French word for the moon) is a concave region in the plane bounded by two circular arcs. Hippocrates of Chios (c. 470-410 BCE) was perhaps the first person to "square" lunes, that is, to show that the area of a certain lune is the same as the area of a certain polygon. He proved the following result concerning a regular hexagon, six lunes, and a circle.

Theorem 3.9.1. *If a regular hexagon is inscribed in a circle and six semicircles constructed on its sides, then the area of the hexagon equals the area of the six lunes plus the area of a circle whose diameter is equal to one of the sides of the hexagon.* See Figure 3.9.1.

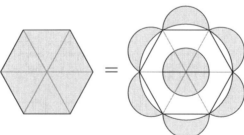

FIGURE 3.9.1

3.9. MISCELLANEOUS EXAMPLES

Proof. See Figure 3.9.2. In the proof we use the fact (known to Hippocrates) that the area of a circle is proportional to the square of its radius, so that the four small gray circles in the second line have the same combined area as the large white circle, so that $4 \cdot \pi(a/2)^2 = \pi a^2$ where a is the side length of the hexagon. ∎

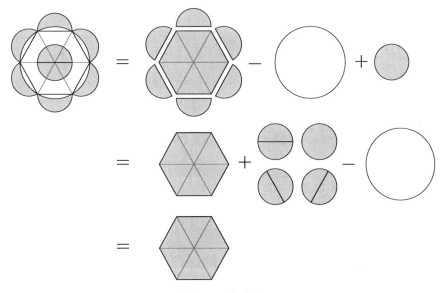

FIGURE 3.9.2

Example 3.9.2. *A hexagonal proof of the Pythagorean theorem.* In Figure 3.9.3a we see the familiar illustration of a right triangle, with squares on the legs and the hypotenuse, enclosed in a rectangle. In Figure 3.9.3b we partition the rectangle into a pair of congruent hexagons, noting that the white regions consist of copies of the triangle. Congruent hexagons have equal area, and each one contains five copies of the triangle, so that when the triangles are deleted, we have $a^2 + b^2 = c^2$. □

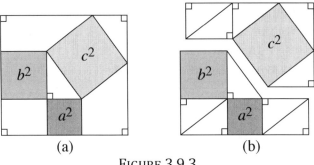

FIGURE 3.9.3

Example 3.9.3. *A hexagonal proof of a Pythagorean-like area theorem.* Let T denote the area of a triangle with side lengths a, b, and c, and let T_a, T_b, and T_c denote the areas of equilateral triangles constructed externally on sides a, b, and c, respectively, as shown in Figure 3.9.4a. If the angle opposite side c measures $120°$, then $T_c = T_a + T_b + T$.

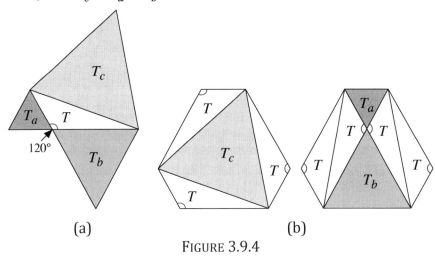

FIGURE 3.9.4

In Figure 3.9.4b we see a pair of congruent equiangular hexagons with areas $T_c + 3T$ and $T_a + T_b + 4T$, from which the result follows. □

Example 3.9.4. *A hexagonal proof of Cassini's identity.* In the L-hexagon, one vertical side is the sum of the other two vertical sides, and the same is true for the horizontal sides. This makes it especially well suited for illustrating identities for *Fibonacci* numbers $\{F_n\}_{n=1}^{\infty} = \{1, 1, 2, 3, 5, 8, 13, 21, ...\}$ defined by $F_1 = F_2 = 1$ and $F_n = F_{n-1} + F_{n-2}$ for $n \geq 3$. Comparing the square of one Fibonacci number to the product of its neighbors in the sequence, e.g., $2 \cdot 5 - 3^2 = +1$, $3 \cdot 8 - 5^2 = -1$, and so on, leads to an identity named for the Italian mathematician Giovanni Domenico Cassini (1625-1712), who discovered it in 1680.

Theorem 3.9.2 (Cassini's identity). *For all $n \geq 2$, $F_{n-1}F_{n+1} - F_n^2 = (-1)^n$.*

Proof. Evaluating the area of the L-hexagon in Figure 3.9.5 in two different ways leads to $F_{n-1}F_{n+1} + F_{n-2}F_n = F_n^2 + F_{n-1}^2$, which is equivalent to

$$F_{n-1}F_{n+1} - F_n^2 = -(F_{n-2}F_n - F_{n-1}^2).$$

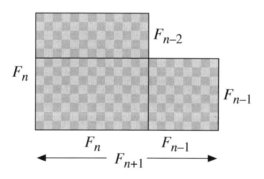

FIGURE 3.9.5

Thus the terms in the sequence $\{F_{n-1}F_{n+1} - F_n^2\}_{n=2}^{\infty}$ have the same magnitude and alternate in sign. So we need only evaluate the base case $n = 2$: $F_1 F_3 - F_2^2 = 1$ so $F_{n-1}F_{n+1} - F_n^2$ is $+1$ when n is even and -1 when n is odd, i.e., $F_{n-1}F_{n+1} - F_n^2 = (-1)^n$. ∎

Example 3.9.5. *The friends and strangers theorem.* The following theorem is somewhat surprising in that an unexpected conclusion follows from rather simple conditions.

Theorem 3.9.3. *At a party of six people, any pair of individuals are either acquainted (friends) or unacquainted (strangers). It* *follows that at least three individuals are either mutual friends or mutual strangers.*

Proof. We use the vertices of a regular hexagon to represent the individuals at the party, and join two vertices by a gray line (an edge or a diagonal) if they are friends, or a black line if they are strangers. See Figure 3.9.6a for an example.

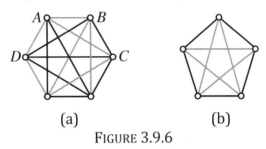

(a)   (b)

FIGURE 3.9.6

We now show that there must be a triangle of lines all gray or all black. Consider one vertex (say A) and the five lines joining it to the other five vertices. At least three of the five must be the same color, say gray. Let B, C, and D be the other endpoints of those lines. If one of the lines BC, BD, or CD is gray, we have a gray triangle. Otherwise BCD is a black triangle. Figure 3.9.6b shows that "six" cannot be replaced by "five" in the theorem. ∎

The James Webb Space Telescope

The primary mirror on the Webb space telescope (launched on December 24, 2021) is 6.5 meters in diameter and consists of 18 hexagonal segments, as illustrated in Figure 3.9.7. Each segment is 1.34 meters in diameter and is made of beryllium coated in gold. A mirror this large is needed in order to capture light from distant galaxies. A hexagonal shape was chosen since the mirror needed a roughly circular shape and was folded upon launch and unfolded once in space.

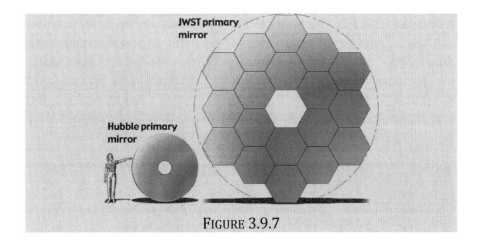

FIGURE 3.9.7

3.10. Hexagons in architecture

While hexagonal structures are not nearly as common as rectangular ones, many do exist. In our first example we have a house and a train station; in the second an adaptation of the yurt, the portable round tent from central Asia.

Example 3.10.1. *Two hexagonal buildings*. On the left in Figure 3.10.1 we see an historic house in Winchester, Virginia, known as the *Hexagon House*. It was constructed in the 1870s, and is now on the National Register of Historic Places.

FIGURE 3.10.1

On the right we see the hexagonal Eindhoven Beukenlaan railway station in the Netherlands, on the day it opened in September 1971. Today it is a snackbar. □

Example 3.10.2. *Hexayurts.* A hexayurt is an inexpensive hexagonal shelter quickly constructed using 4 ft. × 8 ft. sheets of foil-covered plastic foam insulation. On the left in Figure 3.10.2 we see several at the Burning Man Festival in the Black Rock Desert of Nevada in 2010. On the right are sample plans for constructing one. □

 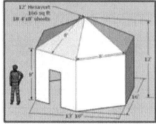

FIGURE 3.10.2

Example 3.10.3. *A hexagonal village.* The village of Grammichele in the province of Catania in Sicily has a rather unusual hexagonal layout of its streets, as shown in Figure 3.10.3. The town was constructed in 1693 after an earthquake destroyed a nearby village. □

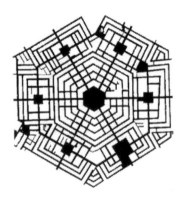

FIGURE 3.10.3

Koch snowflakes

We began this chapter with a quote about and some images of snowflakes. We conclude the chapter with a brief discussion of polygons called *Koch snowflakes*, fractals that approximate the outline of a physical snowflake. These snowflakes first appeared in a 1904 paper by the Swedish mathematician Niels Fabian Helge von Koch (1870-1924). The iterative procedure to draw a Koch snowflake begins with an equilateral triangle. Trisect each side and on each middle third construct an equilateral triangle oriented outwards, erasing line segments common to the old and new triangles. The first iteration produces the outline of a hexagram. Repeating the process produces an infinite sequence of polygons, the first few of which are shown in Figure 3.10.4. The Koch snowflake is the fractal limit of the sequence.

FIGURE 3.10.4

In the limit the number of sides and the perimeter are infinite, but the area is finite, equal to 8/5 of the area of the original equilateral triangle. See [Sandefur, 1996] for details.

3.11. Challenges

3.1 Given an equiangular hexagon with sides of lengths a, b, c, d, e, and f (in that order), show that $a - d = e - b = c - f$.

3.2 Show that the number of calissons that can be placed in a regular hexagonal box such as the one in Figure 3.2.7b is always three times a square.

3.3 In a cyclic hexagon *ABCDEF*, triangle *ACE* is equilateral. Show that the perimeter of *ABCDEF* equals the sum $AD + BE + CF$ of the three long diagonals.

3.4 Prove or disprove the following converse of Theorem 3.3.1. If ABCDEF is a convex hexagon such that $A + C + E = B + D + F = 2\pi$, then ABCDEF is cyclic.

3.5 (i) Find the circumradius R of the cyclic hexagon in Figure 3.3.2a.

(ii) Find the circumradius R of a cyclic hexagon who side lengths are x, x, x, x, y, y ($x \neq y$) in that order. See Figure 3.11.1.

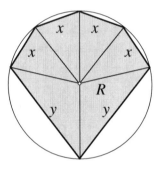

FIGURE 3.11.1

3.6 Let ABCDEF be a convex parahexagon, (i) Show that the area of ABCDEF equals twice the area of $\triangle ACE$. (ii) Let G, H, I, and J be the midpoints of sides BC, CD, EF, and FA, respectively. Show that the area of ABCDEF equals twice the area of GHIJ.

3.7 In Example 3.5.2 we introduce the median triangle associated with a given triangle. If the side lengths of the given triangle are a, b, and c; and the lengths of the medians drawn to a, b, and c are m_a, m_b, and m_c respectively, show that

$$3(a + b + c)/4 \leq m_a + m_b + m_c \leq a + b + c.$$

(Hint. Use the parahexagon in Figure 3.5.3b.)

3.8 Show that every L-polyomino is a reptile.

3.9 Show that the hexagonal "P"-*pentomino* in Figure 3.11.2 is a reptile, and can be dissected into 4, 9, and 16 copies of itself like the L-polyominoes in Figure 3.7.4.

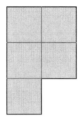

FIGURE 3.11.2

3.10 *Marion Walter's Theorem.* In Figure 3.11.3 we see a triangle whose sides have been trisected. Intersections of the cevians (line segments connecting a vertex to the opposite side of a triangle) drawn to the trisection points yield an irregular hexagon (shaded gray). Show that the area of the hexagon is exactly 1/10 the area of the triangle.

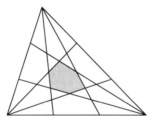

FIGURE 3.11.3

Marion Ilse Walter (1928-2021) was a professor emerita of mathematics at the University of Oregon. The above theorem is often called just *Marion's theorem*.

3.11 In Figure 3.11.4 we again see the familiar illustration of a right triangle with squares on the legs a and b and hypotenuse c, enclosed in an irregular hexagon formed by connecting adjacent vertices of the squares. Show that the area of the hexagon is $(a+b)^2 + c^2$.

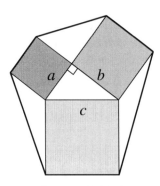

FIGURE 3.11.4

3.12 Show that there exist cyclic hexagons with all sides and diagonals integers. (Hint. Consider a hexagon wherein opposite vertices are endpoints of diameters, so that certain sides and diagonals form right triangles.)

3.13 Is it possible to dissect a regular hexagon into eight congruent quadrilaterals?

3.14 From an arbitrary point within a regular hexagon with side length a and area K draw lines to the vertices creating triangles with areas $T_1, T_2, \ldots, T_6$, as shown in Figure 3.11.5.

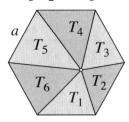

FIGURE 3.11.5

Show that $T_1 + T_3 + T_5 = K/2 = T_2 + T_4 + T_6$.

3.15 *The Babylonian approximation of π.* In 1936 some old Babylonian tablets were discovered near the ancient city of Susa (in present-day Iran). One of the tablets expressed the ratio of the circumference of a regular hexagon to the circumference of its circumcircle as 0;57,36 in the sexagesimal (base 60) number system ($a; b, c = a + \frac{b}{60} + \frac{c}{3600}$). Show that this leads to 25/8 as an approximation of π [Eves, 1983].

3.16 A romantic riddle by an unknown poet goes as follows:

> I am obliged to plant a grove,
> To please the pretty girl I love.
> This curious grove I must compose
> Of nineteen trees in nine straight rows;
> And in each row, five trees must place,
> Or I may never see her face.
> Now readers brave, I'm in no jest,
> Pray lend your aid and do your best.

How should the "poet" plant his grove? (Hint. Begin with a hexagram.)

CHAPTER 4

Heptagons

Since the earliest days of recorded mathematics, the regular heptagon has been virtually relegated to limbo.

L. Bankoff and J. Garfunkel

4.1. Introduction

In [Bankoff and Garfunkel, 1973] the authors justify the above quotation with the following observations.

• Regular heptagons cannot tile the plane (as do equilateral triangles, squares, and regular hexagons), nor do they appear as faces of a Platonic or Archimedean solid, so they are rarely seen in floor tiling or in physical objects.

• Unlike pentagons and hexagons, a regular heptagon cannot be constructed using a compass and straightedge. This is a consequence of the *Gauss-Wantzel theorem* in Section 1.5.

• Regular heptagons do not appear to be associated with an important mathematical constant as are circles (with π), squares (with $\sqrt{2}$) and pentagons (with the golden ratio φ).

Archimedes' original work on the construction of a regular heptagon is lost, but a surviving 9th century Arabic translation by Thâbit ibn Qurra, entitled *Book of the construction of the circle divided into seven equal parts, by Archimedes*, was published in English in the 18th century. However, only the last two of eighteen propositions are related to the heptagon construction [Knorr, 1989].

Fortunately, these facts have not prevented the study of heptagons throughout history. An early example is an approximation of a heptagon partitioned into triangles on an ancient Babylonian tablet in the Musée du Louvre in Paris as seen in Figure 4.1.1.

FIGURE 4.1.1

Heptagonal coins

Coins with polygonal sides enable people with limited sight to distinguish coins by touch. In Figure 4.1.2 we see three heptagonal coins: a 1000 kwacha coin from Zambia, a 50 pence coin from the United Kingdom, and 25 centavos from Brazil.

FIGURE 4.1.2

The Zambian coin is a regular heptagon, while the British coin is a *Reuleaux heptagon*, with curved sides so that the perimeter is a curve of constant width (the width of a convex closed curve is the distance between two parallel lines lying outside the curve but touching its boundary). This enables the coin to function in vending machines. The Brazilian coin is circular with a regular heptagon inscribed in its face.

4.2. Regular heptagons

A *heptagon* is a seven-sided polygon (or a 7-*gon*), and can be convex, concave, or complex, as is the case for heptagonal stars.

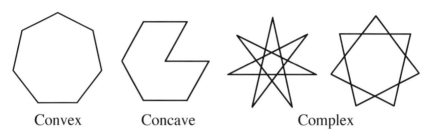

Convex   Concave   Complex

FIGURE 4.2.1

Of interest in this section are the *regular heptagons*, convex heptagons with equal side lengths a and equal vertex angles θ. From Section 1.3 we have the following data for regular heptagons:

vertex angle $\quad \theta = 5\pi/7$

side length $\quad a = 2r\tan\frac{\pi}{7} = 2R\sin\frac{\pi}{7}$

semiperimeter $\quad s = 7a/2$

area $\quad K = \frac{7}{4}a^2 \cot\frac{\pi}{7} = 7r^2 \tan\frac{\pi}{7} = \frac{7}{2}R^2 \sin\frac{2\pi}{7} = rs$

circumradius $\quad R = \frac{a}{2}\csc\frac{\pi}{7}$

inradius $\quad r = \frac{a}{2}\cot\frac{\pi}{7} = R\cos\frac{\pi}{7}$

4.3. The diagonals of a regular heptagon

In Section 1.2 we saw that a regular polygon with k sides has $k(k-3)/2$ diagonals, so the regular heptagon has 14 diagonals. In Figure 4.3.1a we have a regular heptagon *ABCDEFG* with side length a, seven short diagonals (dashed line segments) of length b, and seven long diagonals (solid line segments) of length c.

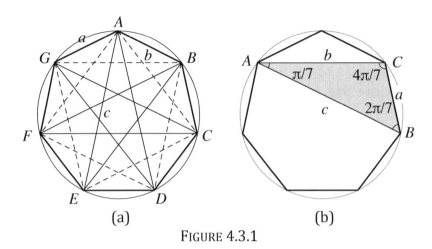

FIGURE 4.3.1

In Figure 4.3.1b we see the *heptagonal triangle ABC* with sides a, b, and c; we have relabeled the vertices to match the opposite sides. We can now evaluate b and c in terms of a. Applying the law of sines to $\triangle ABC$ yields

$$(4.1) \qquad \frac{a}{\sin(\pi/7)} = \frac{b}{\sin(2\pi/7)} = \frac{c}{\sin(4\pi/7)},$$

so that

$$(4.2) \qquad b = 2a\cos(\pi/7) \cong 1.802a \text{ and}$$
$$c = 4a\cos(\pi/7)\cos(2\pi/7) \cong 2.247a.$$

The expressions in (4.2) for b and c also follow from (1.6) after observing that $d_2 = d_3$ since $\sin(3\pi/7) = \sin(4\pi/7)$.

In the following theorem we present four additional relations for a, b, and c.

Theorem 4.3.1. *For the regular heptagon ABCDEFG in Figure 4.3.1a with side a and diagonals b and c, we have* i) $c^2 = bc + a^2$, ii) $b^2 = ac + a^2$, iii) $c^2 = b^2 + ab$, *and* iv) $bc = ac + ab$, *so that* $1/a = 1/b + 1/c$.

Proof. Apply Ptolemy's theorem 1.4.2 to the cyclic quadrilaterals i) BCFG, ii) ABCG, iii) BDEG, and iv) ABCF in Figure 4.3.1a. ∎

In the following corollary we express the ratios b/a and c/a implicit in (4.2) as roots of cubic equations. See Challenge 4.3 for a proof.

Corollary 4.3.2. *With a, b, and c as in Theorem 4.3.1, let $x = b/a$ and $y = c/a$. Then*

$$x^3 - x^2 - 2x + 1 = 0 \quad \text{and} \quad y^3 - 2y^2 - y + 1 = 0.$$

Note that each equation has a negative root, a root in (0,1) and a root larger than 1. So b/a and c/a are the largest roots of the cubic equations.

4.4. The heptagonal triangle

There are a great many interesting identities that one can derive for the angles and sides of the heptagonal triangle ABC in Figure 4.3.1b. We present only a few; for more see the Challenges in Section 4.10 and [Bankoff and Garfunkel, 1973].

Theorem 4.4.1. *For the heptagonal triangle in Figure* 4.3.1b *we have*

(4.3) $\quad \cos A = b/(2a), \cos B = c/(2b), \cos C = -a/(2c),$

and hence

(4.4) $\quad\quad\quad\quad\quad \cos A \cos B \cos C = -1/8.$

Proof. Note that $B = 2A$ and $C = 2B = 4A$. From (4.1) we have $b \sin A = a \sin 2A = 2a \sin A \cos A$ so that $b = 2a \cos A$, and thus $\cos A = b/(2a)$. Similarly $c \sin 2A = b \sin 4A = 2b \sin 2A \cos 2A$ so that $c = 2b \cos 2A = 2b \cos B$, and thus $\cos B = c/(2b)$. Next note that $2C = 8\pi/7 = \pi + A$ so that $\sin 2C = 2 \sin C \cos C = -\sin A$. From (4.1) we have $a \sin C = c \sin A = -2c \sin C \cos C$ so that $a = -2c \cos C$, and thus $\cos C = -a/(2c)$. Multiplying the three cosines in (4.3) yields (4.4). ∎

Corollary 4.4.2. *If A, B, and C are the angles in the heptagonal triangle, then*

(4.5) $$\sin A \sin B \sin C = \sqrt{7}/8.$$

Proof. We show that $64 \sin^2 A \sin^2 B \sin^2 C = 7$. Equation (4.3) and the identities in Theorem 4.3.1 yield
$$4 \sin^2 A = (4a^2 - b^2)/a^2 = (3a^2 - ac)/a^2 = 3 - (c/a),$$
and similarly
$$4 \sin^2 B = 3 - (a/b) \text{ and } 4 \sin^2 C = 3 + (b/c).$$
Hence
$$64 \sin^2 A \sin^2 B \sin^2 C = \left(3 - \frac{c}{a}\right)\left(3 - \frac{a}{b}\right)\left(3 + \frac{b}{c}\right)$$
$$= 27 - 9\left(\frac{c}{a} + \frac{a}{b} - \frac{b}{c}\right) - 3\left(\frac{a}{c} + \frac{b}{a} - \frac{c}{b}\right) + 1.$$
But
$$\frac{c}{a} + \frac{a}{b} - \frac{b}{c} = \frac{a}{b} + \frac{c^2 - ab}{ac} = \frac{a}{b} + \frac{b^2}{ac} = \frac{a}{b} + 1 + \frac{a}{c} = 2$$
and

(4.6) $$\frac{a}{c} + \frac{b}{a} - \frac{c}{b} = \frac{a}{c} + \frac{b^2 - ac}{ab} = \frac{a}{c} + \frac{a^2}{ab} = \frac{a}{c} + \frac{a}{b} = 1.$$

Hence $64 \sin^2 A \sin^2 B \sin^2 C = 27 - 9(2) - 3(1) + 1 = 7.$ ■

Our next theorem relates the sum of the squares of the sides of the heptagonal triangle to its circumradius. First we need the following lemma that applies to arbitrary triangles.

Lemma 4.4.3. *Let A, B, and C denote the angles of an arbitrary triangle. Then*

(4.7) $$\sin^2 A + \sin^2 B + \sin^2 C = 2 + 2 \cos A \cos B \cos C.$$

Proof. When a triangle with angles A, B, and C is scaled so that it can be inscribed in a circle with radius $1/2$, the sides opposite those angles have lengths $\sin A$, $\sin B$, and $\sin C$, respectively, by the extended law of sines. Applying the law of cosines to that triangle yields

$$\sin^2 A = \sin^2 B + \sin^2 C - 2 \sin B \sin C \cos A.$$

Since
$$-\sin B \sin C \cos A = \cos A \cos(B+C) - \cos A \cos B \cos C$$
and
$$\cos A \cos(B+C) = \cos A \cos(\pi - A) = -\cos^2 A = -1 + \sin^2 A,$$
we have
$$\sin^2 A = \sin^2 B + \sin^2 C - 2 + 2\sin^2 A - 2\cos A \cos B \cos C,$$
which is equivalent to (4.7). ∎

Theorem 4.4.4. *If a, b, and c are the sides lengths and R the circumradius of the heptagonal triangle, then*
$$a^2 + b^2 + c^2 = 7R^2.$$

Proof. From the law of sines we have $a = 2R \sin A$, $b = 2R \sin B$, $c = 2R \sin C$, and thus Lemma 4.4.3 and Theorem 4.4.1 yield
$$a^2 + b^2 + c^2 = 4R^2(\sin^2 A + \sin^2 B + \sin^2 C)$$
$$= 4R^2(2 + 2\cos A \cos B \cos C)$$
$$= 4R^2(2 + 2(-1/8)) = 7R^2. \blacksquare$$

For more results about the diagonals of a regular heptagon and the heptagonal triangle, see [Bankoff and Garfunkel, 1973], [Steinbach, 1997], [Steinbach, 2000], [Yiu, 2009], and [Wang, 2019].

4.5. Drawing a regular heptagon

As noted in the introduction to this chapter, it is impossible to use only a straightedge and compass to draw a regular heptagon. But there are a variety of ways to use those two instruments to draw good approximations to a regular heptagon. We present two different ways, one that produces a cyclic irregular heptagon, and another that yields an equilateral irregular heptagon.

Example 4.5.1. *Albrecht Dürer's 1525 approximation.* In his book *Underweysung der Messung mit dem Zirckel und Richtscheyt* (Instructions on Measurement with Compass and Ruler) of 1525, Dürer provides directions for drawing all the "regular" polygons from the triangle to the 16-gon (some exact, some approximate) using only compass and (unmarked) straightedge [Hughes, 2012].

Dürer begins with an equilateral triangle inscribed in a circle of radius 1. Bisecting one side yields a segment of length $\sqrt{3}/2 \cong 0.866025$ as an approximation to the actual side length $2\sin(\pi/7) \cong 0.867767$. See Figure 4.5.1a. This approximation is equivalent to using the inradius of a regular hexagon inscribed in the same circle, as described in Heron of Alexandria's *Metrica* [Bankoff and Garfunkel, 1973]. In Figure 4.5.1b we see a 17th century German print illustrating the construction in the Staats- und Universitätsbibliothek in Dresden. □

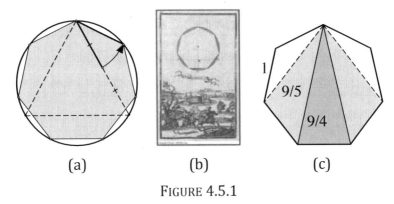

FIGURE 4.5.1

Example 4.5.2. *Approximation by triangulation.* This construction is motivated by the observation that when the side length a equals 1, (4.2) yields the approximations $b = 9/5$ and $c = 9/4$ for the diagonals. These approximations satisfy part iv of Theorem 4.3.1, but not the other three parts. We construct the heptagon using the approximations to the diagonals as shown in Figure 4.5.1c. The result is equilateral but neither equiangular nor cyclic (since the triangles have different circumradii). The area of a regular heptagon with side 1 is approximately

3.6339124, whereas the area of this heptagon is approximately 3.6338631 for a relative error less than 0.0014%. □

For another approximation, see [Running, 1923].

A 3D approximation with compass, straightedge, scissors, and tape.

Given a circle of radius R, draw a circle of radius $8R/7$ divided into eighths on another sheet of paper. Cut out the circle with scissors and discard one eighth. Form the remaining piece into a cone and tape closed. Since the circumference of the base of the cone is the same as that of the original circle, the cone fits on the circle of radius R and divides it into 7 equal parts, inscribing a regular heptagon in the circle.

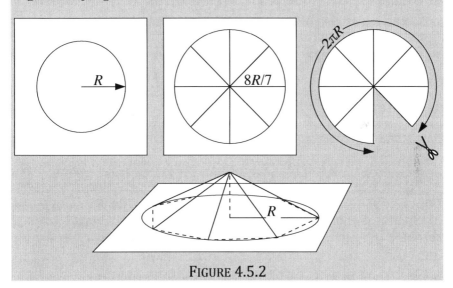

FIGURE 4.5.2

We conclude this section with an exact construction of a regular heptagon using a compass, a straightedge, and the carpenter's square from Section 3.6. It is due to the American mathematician Andrew Mattei Gleason (1921-2008).

Example 4.5.3. *Gleason's construction of a regular heptagon.* Draw a circle of radius 6 centered at the origin of Cartesian plane, and locate points $A(6,0)$, $P(-1,0)$, $Q(-3,0)$, and $R(3,0)$, as shown in Figure 4.5.3. Let $K(0,3\sqrt{3})$ and $L(0,-3\sqrt{3})$ be vertices of equilateral triangles with base QR. With center P draw an arc from K to L and use a carpenter's square to trisect it at S and T. Then the points B and G where the line through S and T intersects the circle are vertices of a regular heptagon $ABCDEFG$. For a proof that the heptagon is regular, see [Gleason, 1988]. □

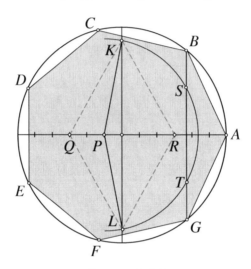

FIGURE 4.5.3

The carpenter's square (or any device that trisects angles) along with compass and straightedge can be used to draw regular n-gons whenever n is a product of a power of 2, a power of 3, and a nonnegative number of distinct *Pierpont primes* (greater than 3) [Gleason, 1988]. Pierpont primes are primes of the form $2^a 3^b + 1$, named for the American mathematician James Pierpont (1866-1938). The first 10 Pierpont primes are 2, 3, 5, 7, 13, 17, 19, 37, 73, 97 (sequence A005109 at oeis.org).

4.6. A neusis construction

The word *neusis*, from a Greek word meaning "inclined towards," is used in mathematics to describe a geometric construction where a *ruler*, or marked straightedge, is used along with the compass. A neusis construction of a true regular

4.6. A NEUSIS CONSTRUCTION

heptagon is possible, and the one we present [Johnson, 1975] illustrates the meaning of the word neusis.

First note that to construct the regular heptagon ABCDEFG in Figure 4.3.1a, we need only construct the triangle ADE. Once we have $\triangle ADE$ we can use perpendicular bisectors of two sides to locate the center of the circumcircle, draw it, and use side DE to draw the other six sides of the heptagon. The following lemma describes the property of $\triangle ADE$ that enables the neusis construction.

Lemma 4.6.1. *Let ADE be an isosceles triangle with base DE of length 1 and equal sides AE and AD of length y. Suppose a point X on side AE is a distance 1 from A and a distance $\sqrt{2}$ from D, as shown in Figure 4.6.1a. Then y equals the length of the long diagonal in a regular heptagon with side length 1.*

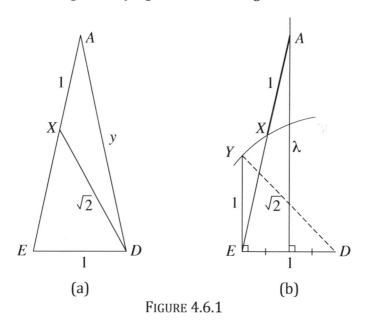

FIGURE 4.6.1

Proof. We apply the law of cosines to $\triangle ADX$ and $\triangle ADE$ in Figure 4.6.1a to obtain

$$2 = 1 + y^2 - 2y \cos A \quad \text{and} \quad 1 = 2y^2 - 2y^2 \cos A.$$

Solving each for $\cos A$ and equating yields the cubic $y^3 - 2y^2 - y + 1 = 0$ in Corollary 4.3.2, whose largest root is the ratio of the long diagonal to the side in a regular heptagon. ∎

To construct $\triangle ADE$ begin by drawing line segment DE of length 1 and line segment EY of length 1 perpendicular to DE, as shown in Figure 4.6.1b. Also draw line λ perpendicular to DE at its midpoint. Draw a circular arc with center D and radius DY. Now perform the neusis step with the ruler. Construct line EA with A on λ and intersecting the circular arc at X so that AX has length 1 (i.e., EA is "inclined towards" λ). Then by the lemma AE is a long diagonal of a heptagon with side length 1, as required.

4.7. Heptagonal tilings

In Sections 2.6 and 3.4 we saw examples of how some convex pentagons and convex hexagons can tile the plane. What about convex heptagons, or, more generally, convex n-gons for $n \geq 7$? The answer is simple: No convex n-gons for $n \geq 7$ can tile the plane [Kerschner, 1969, Niven 1978]. A heuristic explanation goes as follows.

Consider edge-to-edge tilings with convex heptagons. The seven angles of a heptagon total 5π or $900°$, so the average heptagon angle measures $900/7°$. Every vertex of a tiling has $360°$ to be split up among the heptagons meeting there, so on average $360/(900/7) = 2.8$ heptagons meet at each vertex. But since at least 3 heptagons have to meet at each vertex, an average of 2.8 is impossible. Having two heptagons meet at an interior point of an edge, or using polygons with more than 7 sides, only decreases the average number of polygons at each vertex.

However, it is possible to tile the plane with non-convex heptagons. See Challenge 4.1.

4.8. Star heptagons

There are two regular star heptagons, the star polygons {7/2} and {7/3}. When inscribed in a regular heptagon with side length a, the side lengths of {7/2} and {7/3} are the diagonals of length b and c, respectively, of the regular heptagon, as shown in Figure 4.8.1.

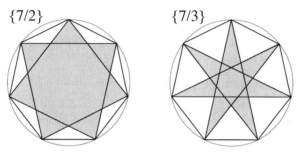

FIGURE 4.8.1

The star heptagon {7/2} is sometimes called simply *the heptagram*, while {7/3} is called *the great heptagram*. Notice that a small regular heptagon and a {7/2} star appear in the center of {7/3}, showing that the sides of a regular heptagon can be extended to meet forming a {7/2} star, and further extended to meet a second time forming a {7/3} star. Some data for {7/2} and {7/3} appear below. The circumradius R for both {7/2} and {7/3} is the one for the regular heptagon, $R = (a/2)\csc(\pi/7)$.

	{7/2}	{7/3}
vertex angle	$3\pi/7$	$\pi/7$
side length	$b = 2a\cos\dfrac{\pi}{7}$	$c = 4a\cos\dfrac{\pi}{7}\cos\dfrac{2\pi}{7}$

For the inradii $r_{\{7/2\}}$ and $r_{\{7/3\}}$ of the two stars, we have

$$r_{\{7/2\}} = \frac{b}{2}\cot\frac{2\pi}{7} = R\cos\frac{2\pi}{7} \quad \text{and} \quad r_{\{7/3\}} = \frac{c}{2}\cot\frac{3\pi}{7} = R\cos\frac{3\pi}{7}.$$

Using (1.6) and (1.7) to evaluate the areas $[\{7/2\}, a]$ and $[\{7/3\}, a]$ of {7/2} and {7/3}, respectively, yields

and
$$[\{7/2\}, a] = \frac{7}{2}a^2 \cot\frac{2\pi}{7} \cong 2.79116a^2$$

$$[\{7/3\}, a] = \frac{7}{4}a^2 \left(\cot\frac{\pi}{7} - \tan\frac{2\pi}{7}\right) \cong 1.43948a^2.$$

2.9. Heptagons in architecture

In this section we present several examples employing the regular heptagon in architecture. Examples are rare, perhaps due to the concerns expressed in the introduction to this chapter. However, the number 7 frequently appears in religious and mythological symbolism, so perhaps it is not surprising to see seven-sided polygons used in certain situations.

Example 4.9.1. *The Mausoleum of Prince Ernst in Stadthagen, Germany.* In Figure 4.9.1a we see the mausoleum, in the shape of a domed heptagonal prism, attached to the parish church in Stadthagen. Figure 4.9.1b is a view inside looking upwards toward the dome. □

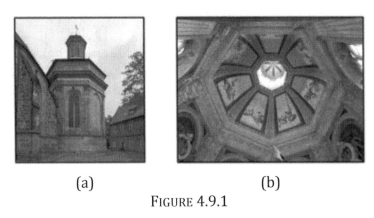

(a) (b)

FIGURE 4.9.1

Example 4.9.2. *The baldachin in the Sagrada Família in Barcelona.* A *baldachin*, or baldaquin, is an ornamental structure, resembling a canopy, used over an altar in a church. In Figure 4.9.2a we see one in the shape of a regular heptagon over the high altar in the Basilica de la Sagrada Família in Barcelona, Spain. □

4.9. HEPTAGONS IN ARCHITECTURE

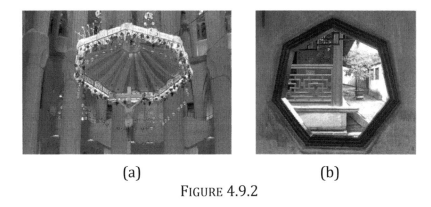

(a)   (b)

FIGURE 4.9.2

Example 4.9.3. *The Yuyuan Garden in Shanghai, China*. In Figure 4.9.2b we see a heptagonal window in this 16th century large Chinese garden in Shanghai's Old City. □

Crockett Johnson's mathematical art

Crockett Johnson was the pen name of the American children's book illustrator and cartoonist David Johnson Liesk (1906-1975). In the last decade of his life he became interested in mathematics and created over 100 paintings with mathematical themes, 80 of which now reside in the National Museum of American History at the Smithsonian Institution in Washington.

FIGURE 4.9.3

On the left in Figure 4.9.3 is the cover of the September 2009 issue of *Math Horizons*, with a figure for the neusis construction of a regular heptagon in Section 4.6 and Johnson's 1975 painting *Construction of the Heptagon*. On the right is a drawing of the painting by the authors.

4.10. Challenges

4.1 While regular heptagons do not tile the plane, show that some non-convex ones do. (Hint: Consider 5/6 of a regular hexagon.)

4.2 Show that in a regular heptagon, the harmonic mean of the lengths of the short and long diagonals equals twice the side length.

4.3 Prove Corollary 4.3.2.

4.4 If a, b, c, and R denote the side length, short diagonal, long diagonal, and circumradius of a regular heptagon, show that

(i) $b + c - a = \sqrt{7}R$, (ii) $\dfrac{b^2}{a^2} + \dfrac{c^2}{b^2} + \dfrac{a^2}{c^2} = 5$, and

(iii) $abc = \sqrt{7}R^3$.

4.5 Let h_a, h_b, and h_c denote the lengths of the altitudes to sides a, b, and c of the heptagonal triangle ABC in Figure 4.3.1b. Show that $h_a = h_b + h_c$.

4.6 Show that the area of the heptagonal triangle ABC in Figure 4.3.1b is $\sqrt{7}R^2/4$, where R is its circumradius.

4.7 Let A, B, C, and D be consecutive vertices of a regular heptagon. Let X denote the intersection of AC and BD, as shown in Figure 4.10.1. Show that $AB + AX = AD$.

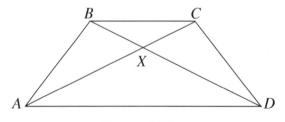

Figure 4.10.1

4.8 Show that the distance *PQ* from the midpoint *P* of side *AB* of a regular convex heptagon *ABCDEFG*, inscribed in a circle, to the midpoint *Q* of the radius perpendicular to *BC*, is equal to half the side length of a square inscribed in the circle.

4.9 A heptagon *ABCDEFG* is inscribed in a circle and three of its angles each equal 120°. Show that the heptagon has two sides of equal length.

4.10 Let *ABCDEFG* be a regular heptagon, and let *P* be any point on its circumcircle between *G* and *A*. Show that $PA + PC + PE + PG = PB + PD + PF$. (This is a companion result to Theorems 2.2.2 and 3.2.1 for pentagons and hexagons.)

4.11 Among the many puzzles created by Sam Loyd (1841-1911) one finds "the sedan chair," shaped like a concave heptagon as shown in Figure 4.10.2 with an illustration from [Loyd, 1914] and the heptagonal side of the chair.

 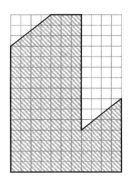

FIGURE 4.10.2

Loyd's challenge is "to cut the sedan chair into the fewest possible pieces which will fit together and form a perfect square." (Hint. Two pieces suffice.)

CHAPTER 5

Octagons

The octagon was seen symbolically as the "intermediary"— the connecting shape— between the circle and the square. It was suggested that the octagon is a circle attempting to become a square, and a square attempting to become a circle.

Mark A. Reynolds
The Octagon in Leonardo's Drawings.

5.1. Introduction

Octagons are eight-sided polygons—or 8-gons—and, like the polygons in preceding chapters, can be convex, concave, complex, compound, and skew. In Figure 5.1.1 we see representatives of those five types of octagons.

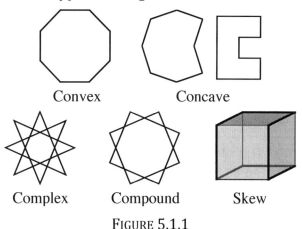

FIGURE 5.1.1

The convex octagon in the above figure is a regular octagon; the subject of the next section. We encounter irregular convex octagons in Section 5.3 and complex and compound (star) octagons in Section 5.4. The path visiting the eight vertices of a cube is an equilateral and equiangular skew octagon.

Octagonal logos and signs

Octagons—regular, irregular, and star—are, like pentagons and hexagons, often employed as logos, signs, and military insignia. Below are a few examples.

FIGURE 5.1.2

5.2. Regular octagons

Inscribing a regular octagon in a circle is almost as easy as it was to inscribe a regular hexagon in Section 3.2. After drawing a pair of perpendicular diagonals, bisect adjacent right angles at the center and extend the bisectors to the circle. Then we have eight points on the circle that are the vertices of a regular octagon, as shown in Figure 5.2.1.

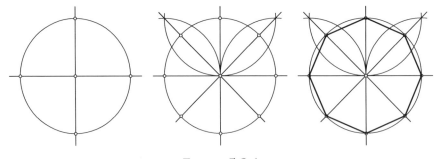

FIGURE 5.2.1

See Challenge 5.5 for a procedure for inscribing a regular octagon in a square.

8.2. REGULAR OCTAGONS

Objects with regular octagonal shapes

Many physical objects with an octagonal shape employ regular octagons, such as the ones in Figure 5.2.2. From left to right, we have an American 50-dollar gold piece commemorating the Pacific-Panama International Exposition in San Francisco in 1915; an octagonal window in the Lingering Garden in Suzhou, China; and a 17th century Hungarian silver dish, currently in the Metropolitan Museum of Art in New York City.

FIGURE 5.2.2

On the right is perhaps the most frequently observed regular octagon, the ubiquitous road stop sign. The one above is from Nepal; note the absence of any text as its octagonal shape and color (red) are sufficient to convey its meaning.

From Section 1.3 we have the following data for a regular octagon with equal side lengths a and vertex angles θ:

vertex angle	$\theta = 3\pi/4 = 135°$
side length	$a = 2r \tan \frac{\pi}{8} = 2R \sin \frac{\pi}{8}$
semiperimeter	$s = 4a$
area	$K = 2a^2 \cot \frac{\pi}{8} = 8r^2 \tan \frac{\pi}{8} = 4R^2 \sin \frac{\pi}{4}$
circumradius	$R = \frac{a}{2} \csc \frac{\pi}{8}$
inradius	$r = \frac{a}{2} \cot \frac{\pi}{8} = R \cos \frac{\pi}{8}$

To make use of the above data, we need to evaluate the trigonometric functions of $\pi/8$. This is most easily done by first finding the lengths of the 20 diagonals of a regular octagon.

There are 8 short diagonals of length d_1, 8 medium diagonals of length d_2, and 4 long diagonals of length d_3, as seen in Figure 5.2.3a.

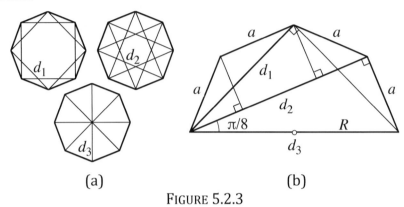

FIGURE 5.2.3

In Figure 5.2.3b we see that $d_3 = 2R$, $d_1 = \sqrt{2}R$, and two expressions for d_2, hence $d_2 = a\cot(\pi/8) = a(\sqrt{2}+1)$. Thus $\cot(\pi/8) = \sqrt{2}+1$ and elementary trigonometric identities now yield

$$\sin(\pi/8) = \tfrac{1}{2}\sqrt{2-\sqrt{2}}, \quad \cos(\pi/8) = \tfrac{1}{2}\sqrt{2+\sqrt{2}},$$
$$\tan(\pi/8) = \sqrt{2}-1, \quad \cot(\pi/8) = \sqrt{2}+1,$$
$$\sec(\pi/8) = \sqrt{4-2\sqrt{2}}, \quad \csc(\pi/8) = \sqrt{4+2\sqrt{2}}.$$

Since $R = (a/2)\sqrt{4+2\sqrt{2}}$ we now have $d_3 = a\sqrt{4+2\sqrt{2}}$ and $d_1 = a\sqrt{2+\sqrt{2}}$. We also have the following expressions for some of the data for the regular octagon.

side length $\quad a = 2r(\sqrt{2}-1) = R\sqrt{2-\sqrt{2}}$

area $\quad K = 2a^2(\sqrt{2}+1) = 8r^2(\sqrt{2}-1) = 2\sqrt{2}R^2$

circumradius $\quad R = \tfrac{a}{2}\sqrt{4+2\sqrt{2}}$

inradius $\quad r = \tfrac{a}{2}(\sqrt{2}+1) = \tfrac{1}{2}R\sqrt{2+\sqrt{2}}$

8.2. REGULAR OCTAGONS

In the next theorem, we express the area of a regular octagon in terms of the diagonals in two ways.

Theorem 5.2.1. *Let a be the side length, K the area, and d_1, d_2, and d_3 the lengths of the diagonals of a regular octagon, as in Figure 5.2.3. Then $K = 2ad_2 = d_1 d_3$.*

Proof. See Figure 5.2.4. ∎

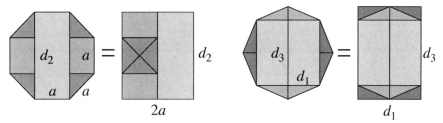

FIGURE 5.2.4

Example 5.2.1. *A hexagon-octagon approximation of π.* In Figure 5.2.5 we see a unit circle with area π, its circumscribed regular hexagon with area $2\sqrt{3}$, and its inscribed regular octagon with area $2\sqrt{2}$. The figure suggests that an approximation to π consists of the arithmetic mean of the two areas, i.e., $\pi \approx \sqrt{2} + \sqrt{3}$.

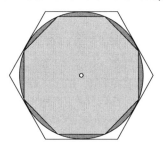

FIGURE 5.2.5

The approximation is actually quite good since $\sqrt{2} + \sqrt{3} \approx 3.14626$, for a relative error of approximately 0.1487%. □

Example 5.2.2. *A geometric series.* When $n \geq 4$ the geometric series

$$\frac{1}{n} + \frac{1}{n^2} + \frac{1}{n^3} + \cdots = \frac{1}{n-1}$$

can be represented by a regular $(n-1)$-gon with side length 1 partitioned into n congruent trapezoids and a similar regular $(n-1)$-gon whose edge length is $1/\sqrt{n}$. In Figure 5.2.6a, we see the $n=9$ case with a regular octagon, where the side length of the inner octagon is $1/3$ [Tanton, 2008].

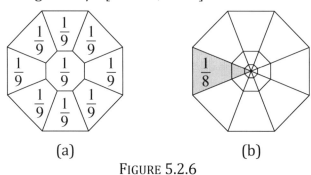

(a) (b)

FIGURE 5.2.6

When we iterate the process, the sequence of trapezoids in the gray region of Figure 5.2.6b illustrates $1/9 + 1/81 + 1/729 + \cdots = 1/8$. □

Example 5.2.3. *The silver ratio.* In Section 2.2 we encountered the golden ratio φ, the positive root of $x^2 - x - 1 = 0$, as the ratio of the length of a diagonal to the length of a side in a regular pentagon. It is the first in a sequence of *metallic ratios*, the positive roots φ_n of $x^2 - nx - 1 = 0$ for n a positive integer (note that $\varphi_1 = \varphi$). When $n = 2$ we have the *silver ratio* $\varphi_2 = \sqrt{2} + 1$. Analogous to the situation with φ and a regular pentagon, φ_2 is the ratio of the length d_2 of a medium length diagonal to the length a of a side in a regular octagon. Analogous to a golden rectangle is a *silver rectangle*, one similar to the rectangle with sides a and d_2 in a regular octagon, as seen shaded light gray in the leftmost octagon in Figure 5.2.4. □

Example 5.2.3. *The Cordovan proportion.* Another constant associated with a regular octagon is the *Cordovan proportion, c,* introduced in 1973 by the Spanish architect Rafael de la Hoz (1924-2000), who discovered that this ratio appears in several Moorish buildings such as the Mezquita in Córdoba, Spain [de la Hoz, 1996]. The constant c is the ratio of the circumradius R to the side length a of a regular octagon; i.e.,

$$c = \frac{R}{a} = \frac{1}{\sqrt{2-\sqrt{2}}} = \sqrt{1 + \frac{\sqrt{2}}{2}} \approx 1.306563.$$

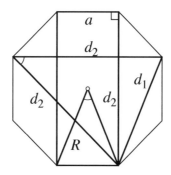

FIGURE 5.2.7

It is also equal to d_2/d_1. Geometrically it equals the ratio of the long side to the short side of an isosceles triangle with sides R, R, a, or d_2, d_2, d_1 in a regular octagon, as illustrated in Figure 5.2.7 (along with an $a \times d_2$ silver rectangle). These types of triangles are called *Cordovan triangles*. The silver ratio and the Cordovan proportion are related by various identities such as $2c^2 = 1 + \varphi_2$ and $\varphi_2 = \sqrt{2}c^2$.

See [Redondo Buitrago and Reyes Iglesias, 2008] for more on the Cordovan proportion. □

Example 5.2.4. *Tiling with octagons.* As noted in Section 4.7, it is impossible to tile the plane with copies of a convex octagon. However, as we noted in Example 2.6.3, there is a semiregular tiling of the plane consisting of regular octagons and squares that can be modified to produce the tiling with Type 10 irregular pentagons. This semiregular tiling has long been used to tile floors. See Figure 5.2.8 for an example from the ancient Roman city Aquincum near the modern city of Budapest. □

FIGURE 5.2.8

5.3. General convex octagons

We present two theorems and two examples concerning convex octagons. In the first theorem we find the area of a particular octagon within a parallelogram, and in the second we present an angle-sum property of cyclic octagons. The examples concern an octagon employed by the ancient Egyptians to approximate the area of a circle and a drawing of a four-dimensional object.

Theorem 5.3.1. *If one joins each vertex of a parallelogram to the midpoints of the two opposite sides, the lines form an irregular convex octagon whose area equals 1/6 of the area of the parallelogram.* See Figure 5.3.1a.

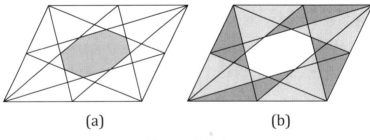

(a)   (b)

FIGURE 5.3.1

Proof. Assume the area of the parallelogram is 1. We show that the area of the shaded octagon in Figure 5.3.1a is 1/6. First recolor the parallelogram as shown in Figure 5.3.1b. Each of the four light gray triangles has area 1/8, since the area of each one is one-half the area of a parallelogram with area 1/4, as shown in Figure 5.3.2a.

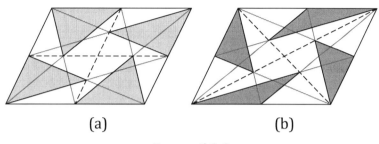

(a) (b)

FIGURE 5.3.2

In Figure 5.3.2b each diagonal partitions the parallelogram into pairs of triangles with area 1/2, and the medians in each of those triangles partitions the triangle into 6 smaller triangles with equal area. Thus each of the four dark gray triangles has area 1/12, and hence the area of the shaded octagon in Figure 5.3.1a is

$$1 - 4 \cdot \frac{1}{8} - 4 \cdot \frac{1}{12} = 1 - \frac{1}{2} - \frac{1}{3} = \frac{1}{6}. \blacksquare$$

The octagon in Theorem 5.3.1 has opposite sides parallel and of equal length. Convex 2n-gons with these properties are known as *zonogons* and are the subject of Section 7.4.

Theorem 5.3.2. *In the cyclic octagon ABCDEFGH each set of four nonadjacent angles sums to* 3π, *i.e.,* $A + C + E + G = B + D + F + H = 3\pi$.

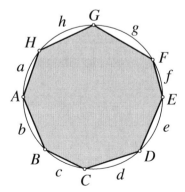

FIGURE 5.3.3

Proof. See Figure 5.3.3, where we assume that the circumradius of *ABCDEFGH* is 1 and the lengths of the eight arcs of the circumference are a, b, c, d, e, f, g, and h as shown. Then $A = (2\pi - a - b)/2$, $C = (2\pi - c - d)/2$, $E = (2\pi - e - f)/2$, and $G = (2\pi - g - h)/2$. Summing we have $A + C + E + G = (8\pi - 2\pi)/2 = 3\pi$. Since the sum of all eight angles is 6π we also have $B + D + F + H = 3\pi$. ∎

The above theorem (with a similar proof) actually holds for any cyclic $2n$-gon, $n \geq 2$, where the sum of n nonadjacent angles is $(n-1)\pi$. See Example 7.5.2 in Section 7.5, where we discuss many additional properties of cyclic polygons.

Example 5.3.1. *An octagon in the Rhind papyrus.* The Rhind papyrus, an ancient Egyptian scroll, was written in about 1650 BCE and currently resides in the British Museum in London. It contains 85 mathematical problems in arithmetic, algebra, and geometry. Problems 41, 42, 43, and 48 concern the area of a circle, and in each the area is given as the area of a square whose side is 8/9 the diameter of the circle. Why 8/9? The answer may be in the text for Problem 48, where we see an irregular octagon inscribed in a square in Figure 5.3.4a.

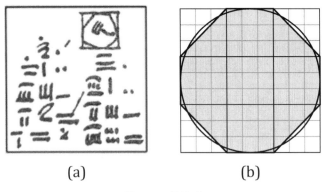

(a)   (b)

FIGURE 5.3.4

Let us assume that the figure represents a square with its sides trisected and then the corners removed, as shown in Figure 5.3.4b along with an inscribed circle. Then it appears that the area of the octagon approximates to the area of the circle. If the diameter of the circle and the side of the square are each 9, then the area of the octagon is $81 - 4(9/2) = 63$, so that side of a square with the same area as the circle would be $\sqrt{63}$, or approximately $\sqrt{64} = 8$. Thus the side of a square, with the same area as the circle, is about 8/9 the diameter of the circle. The resulting approximation of π is $[2(8/9)]^2 \cong 3.160494$. □

Example 5.3.2. *Drawing a tesseract.* One way to draw a picture of a cube is to draw two overlapping squares and connect corresponding vertices. What one has really drawn is an equilateral hexagon with some interior line segments. A *tesseract* or *hypercube* is a four-dimensional analog of an ordinary three-dimensional cube. To draw one, draw two overlapping cubes and connect corresponding vertices with line segments, as shown in Figure 5.3.5a. The result is a regular octagon with some interior line segments. The tesseract has 16 vertices, 32 edges, 24 square faces, and 8 cubic hyperfaces.

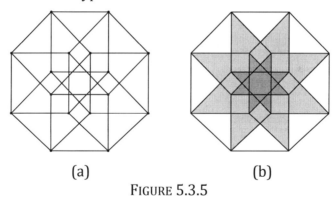

(a)　　　　　　　　(b)

FIGURE 5.3.5

In the drawing of the tesseract as a planar octagon in Figure 5.3.5b we see two eight-pointed stars, called *octagrams*, which we discuss in the next section. □

5.4. Star octagons

There are two regular star octagons, the *Star of Lakshmi* {8/2} and the *octagram* {8/3}. The Star of Lakshmi is a compound star, consisting of two concentric congruent squares with sides inclined at 45° angles, i.e., {8/2} = 2{4}. The Star represents Ashtalakshmi, the eight forms of wealth in Hinduism.

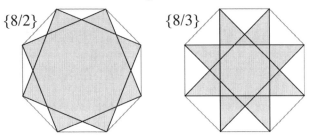

FIGURE 5.4.1

The octagram {8/3} is a unicursal regular star octagon. Notice that a small regular octagon and an {8/2} star appear in the center of {8/3}, showing that the sides of a regular octagon can be extended to meet forming a {8/2} star and further extended to meet a second time forming a {8/3} star. Some data for {8/2} and {8/3} appear below, where a is the side length of the circumscribing regular octagon.

As noted in Section 1.6, the side lengths of {8/2} and {8/3} are the diagonals d_1 and d_2, respectively, of the enclosing regular octagon.

	{8/2}	{8/3}
vertex angle	90°	45°
side length	$d_1 = a\sqrt{2+\sqrt{2}}$	$d_2 = a(\sqrt{2}+1)$
area	$4a^2$	$2a^2\sqrt{2}$
inradius	$r_{\{8/2\}} = \frac{d_1}{2} = \frac{R\sqrt{2}}{2}$	$r_{\{8/3\}} = \frac{a}{2} = \frac{R}{2}\sqrt{2-\sqrt{2}}$

The circumradius R for both {8/2} and {8/3} is the same as for the regular octagon, i.e., $R = (a/2)\sqrt{4+2\sqrt{2}}$.

Example 5.4.1. *The area formulas for regular star octagons.* In Figure 5.4.2 we illustrate the area formulas for {8/2} and {8/3}. In Figure 5.4.2a we replace each of the eight isosceles right triangular points of the star with a triangle having the same base and an altitude of equal length (note that the dashed line segments are parallel to the bases of the triangles), hence the same area, to form a square with side $2a$.

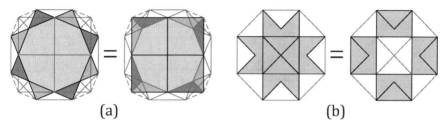

(a)   (b)

FIGURE 5.4.2

In Figure 5.4.2b we show that the area of an octagram $\{8/3\}$ equals the area of four $a \times a/\sqrt{2}$ rectangles, i.e., $2a^2\sqrt{2}$. □

The octagram as a compass rose

An octagram colored as shown in Figure 5.4.3 is called a *compass rose*, and is used on maps and nautical charts to indicate direction. One can be found on the marker for the *point zéro* of France, a spot near the Cathédrale Notre-Dame de Paris from which all highway distances in France are measured.

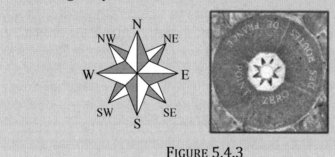

FIGURE 5.4.3

5.5. Octagons in space

Like pentagons and hexagons, a variety of solids have regular octagons and other regular polygons as faces. In Figure 5.5.1a we see a *truncated cube*, with eight equilateral triangular faces and six regular octagonal faces. Figure 5.5.1b shows a sculpture by

Karl-Ludwig Schmaltz of a truncated cube in Würzburg, Germany.

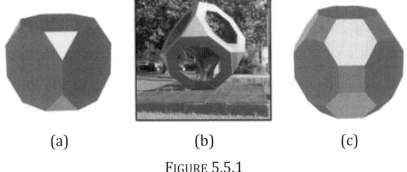

(a) (b) (c)

FIGURE 5.5.1

In Figure 5.5.1c we see a *truncated cuboctahedron*, with 12 square faces, eight regular hexagonal faces, and eight regular octagonal faces.

5.6. Octagons in architecture

In 1848 Orson Squire Fowler published *The Octagon House: A Home for All*, and within a decade over 1000 eight-sided homes (as well as schoolhouses, barns, and businesses) had been built in the United States. Fowler claimed that since regular octagons have a greater area for a given perimeter than do squares, octagon houses would be cheaper to build and easier to heat and cool than square ones.

In the examples below we illustrate some octagonal houses, octagonal schoolhouses, octagonal floor plans in early houses of worship, and the use of octagons in urban planning.

Example 5.6.1. *Two octagonal houses*. Two attractive ones built using some of Fowler's ideas are shown in Figure 5.6.1. On the left we see the *John Richards Octagon House* in Watertown, Wisconsin. It was built in 1854 and placed on the National Register of Historic Places in 1971.

FIGURE 5.6.1

On the right we have the *McElroy Octagon House* in San Francisco, California. It was built in 1861 and placed on the National Register of Historic Places in 1972. Both houses are open to the public. □

Example 5.6.2. *Octagonal schoolhouses.* A number of small octagonal schoolhouses were constructed in Pennsylvania in the early 18th century. On the left in Figure 5.6.2 is the *Wrightstown Octagonal Schoolhouse* in Bucks County. This one-room school was constructed in 1804 and placed on the National Register of Historic Places in 2007.

FIGURE 5.6.2

On the right is the Birmingham Friends School, located in Chester County. This one-room school was constructed in 1819 and placed on the National Register of Historic Places in 1971. □

Example 5.6.3. *Octagons in houses of worship.* The use of octagons in architecture is not a recent development as the preceding examples in this section might indicate. The *Dome of the Rock* in Jerusalem was first constructed in the 7th century. It is the world's oldest existing Islamic monument and has a regular octagonal floor plan as illustrated on the left in Figure 5.6.3.

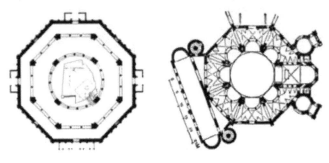

FIGURE 5.6.3

The *Basilica of San Vitale* in Ravenna, Italy was constructed in the 6th century, and also has an regular octagonal floor plan, as shown on the right in Figure 5.6.3. Both the Dome of the Rock and the Basilica of San Vitale are UNESCO World Heritage Sites. □

Example 5.6.4. *Octagons in urban design*. The Catalan urban planner Ildefons Cerdà i Sunyer (1815-1876) designed the extension of Barcelona called the Eixample, which is now part of central Barcelona. His plan, using irregular octagonal city blocks and square street intersections as shown in Figure 5.6.4, is a modified version of the plane tiling using regular octagons and squares in Figure 5.2.8. □

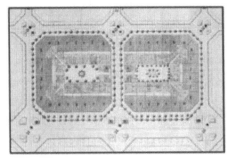

FIGURE 5.6.4

8.6. OCTAGONS IN ARCHITECTURE

Leonardo da Vinci's use of octagons

The Italian genius Leonardo da Vinci (1452-1519) employed a variety of polygons in his architectural drawings. In Figure 5.6.5 we see a floor plan for a temple from a manuscript now in the Bibliothèque de l'Institut de France in Paris. The plan utilizes a regular octagon to locate the chapels surrounding the central space, which contains an octagram in its center. For details, see [Reynolds, 2008].

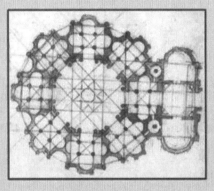

FIGURE 5.6.5

5.7. Challenges

5.1 While regular convex octagons do not tile the plane, show that some concave ones do. (Hint. See Figure 5.1.1.)

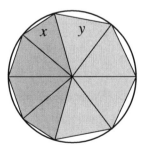

FIGURE 5.7.1

5.2 *A Putnam octagon.* Find the area of a convex octagon that is inscribed in a circle and has four consecutive sides of length x units and the remaining four sides of length y units. See

Figure 5.7.1. This problem (with $x = 2$ and $y = 3$) was problem B-1 on the 1978 William Lowell Putnam Mathematical Competition.

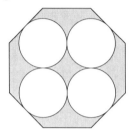

FIGURE 5.7.2

5.3 Inscribe four congruent circles in a regular octagon, as shown in Figure 5.7.2. Show that the diameter of each circle equals the side length of the octagon.

5.4 Let *ABCDEFGH* be a regular octagon inscribed in a unit circle. Show that

$$AB \cdot AC \cdot AD \cdot AE \cdot AF \cdot AG \cdot AH = 8.$$

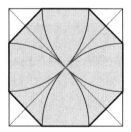

FIGURE 5.7.3

5.5 Figure 5.7.3 illustrates a procedure for inscribing an octagon in a square. Draw circular arcs with radii one-half the diagonal and centers at the vertices of the square. The intersections of the arcs with the sides of the square are the vertices of an octagon. Show that the octagon is regular.

5.6 Suppose two concentric squares with parallel sides have areas in the ratio 2:1. Show that segments drawn through the vertices of the smaller square perpendicular to the diagonals form with the segments of the larger square a regular octagon. See Figure 5.7.4.

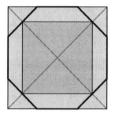

FIGURE 5.7.4

5.7 *Another Putnam octagon.* The octagon *ABCDEFGH* is inscribed in a circle with vertices around the circumference in the given order. Given that the polygon *ACEG* is a square with area 5 and the polygon *BDFH* is a rectangle of area 4, find the maximum possible area of the octagon. (This was problem A-3 on the 2000 William Lowell Putnam Mathematical Competition.)

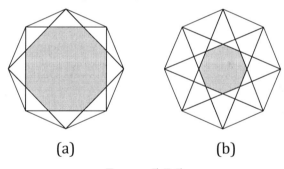

(a)   (b)

FIGURE 5.7.5

5.8 In Figure 5.7.5 we see the inner octagons (shaded gray) in the Star of Lakshmi and the octagram. In each case what fraction of the circumscribing regular octagon is gray?

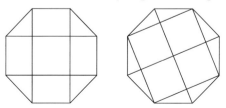

FIGURE 5.7.6

5.9 Show that a regular octagon can be dissected into four congruent hexagons. (Hint. Exploit the symmetry of the octagon by drawing a few diagonals as shown in Figure 5.7.6.)

5.10 In Figure 5.7.7 we have two octagons inscribed in congruent squares. The octagon on the left is formed by joining the midpoints of the sides to the vertices. On the right the trisection points of the sides are used instead. Show that the two octagons are similar and find the ratio of their areas.

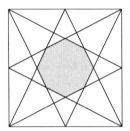 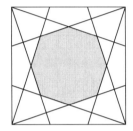

FIGURE 5.7.7

5.11 Use the ratio of the side length and second diagonal of a regular octagon to show that $\sqrt{2}$ is irrational. (Hint. Note that $\sqrt{2}$ is irrational if and only if $\sqrt{2}+1$ is irrational. Assume $\sqrt{2}+1$ is rational, i.e., $\sqrt{2}+1 = m/n$ where m and n are integers with $0 < n < m$ and the fraction is in lowest terms. Then draw the octagon in Figure 5.7.8 with side length n to obtain a contradiction.)

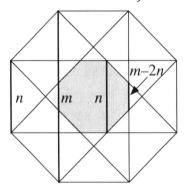

FIGURE 5.7.8

CHAPTER 6

Many-sided Polygons

In 1796, at the age of 19 Gauss proved that a regular 17-gon is constructible with a compass and a straightedge. With this result Gauss became the first person to make headway on the 2,200-year-old geometry problem of determining which regular n-gons are constructible. In fact, he later solved the problem completely [Bliss et al., 2013].

6.1. Introduction

It is said that when Carl Friedrich Gauss (1777–1855) discovered that a regular *heptadecagon*, a polygon with 17 sides, could be constructed with only a compass and straightedge, he decided to devote his life to mathematics. See Figure 6.1.1 for an image of a regular 17-gon.

FIGURE 6.1.1

Later in his life Gauss expressed the wish to have a regular 17-gon inscribed on his tombstone. That wish was never carried out, as the stonemason thought that such a figure would be indistinguishable from a circle [Boyer, 1968; Eves, 1983]. Perhaps the reader agrees.

In Sections 1.5 and 4.1 we briefly mentioned the *Gauss-Wantzel theorem*, concerning which regular polygons can be constructed with compass and straightedge. The theorem states that a regular n-gon can be so constructed if and only if n is product of a power of 2 and distinct *Fermat primes*, primes of the form $2^k + 1$ where k is a power of 2. To date just five Fermat primes are known: 3, 5, 17, 257, and 65537, and it is doubtful others exist. It is easy to draw equilateral triangles, and we have seen in Chapter 2 how to draw regular pentagons. We consider the 17-, 257-, and 65537-gons in Sections 6.6 and 6.8.

Polygons with many sides have played important roles in the history of mathematics. The reader is no doubt familiar with some of the work of Archimedes. Archimedes obtained bounds on π by drawing a circle, inscribing and circumscribing hexagons, and computing their perimeters. He then doubled and redoubled the number of sides until he had polygons with 96 sides, proving that $3\frac{10}{71} < \pi < 3\frac{1}{7}$. In Section 6.7 we discuss the 24-, 48-, and 96-gons and this inequality.

In this chapter we consider some many-sided n-gons in order of increasing n beginning with 9. When π/n radians is not an integer number of degrees, we may not include all the data about side length, area, etc.; in that case please refer to Section 1.3.

Many-sided polygonal coins

In Chapter 4 we saw several examples of seven-sided coins. Coins based on polygons with 9, 11, and 13 sides have also been minted. In Figure 6.1.2 we see three examples. On the left is a 9-sided five euro coin from Austria. In the center we have an 11-sided one dollar coin from Canada. On the right is a 13-sided 20 korun coin from the Czech Republic.

FIGURE 6.1.2

In each case the shape is actually a Reuleaux polygon like the heptagonal British coin in Figure 4.1.2. See Figure 6.5.9 for images of a 12-sided coin.

6.2. Nonagons

A *nonagon* is a nine-sided polygon. It is also called an *enneagon*, using the Greek prefix for nine rather than the Latin non-. One would expect the word enneagon to be preferred, since the words pentagon, hexagon, heptagon, and so on, use Greek prefixes rather than Latin ones such as quint-, sex- and sept-. But nonagon appears to be the more common name for a 9-gon.

From Section 1.3 we have the following data for a regular nonagon with equal side lengths a and vertex angles θ, where we express $\pi/9$ radians as $20°$.

vertex angle $\quad \theta = 7\pi/9 = 140°$

side length $\quad a = 2r \tan 20° = 2R \sin 20°$

semiperimeter $\quad s = 9a/2$

area $\quad K = \dfrac{9}{4}a^2 \cot 20° = 9r^2 \tan 20° = \dfrac{9}{2}R^2 \sin 40°$

circumradius $\quad R = \dfrac{a}{2} \csc 20°$

inradius $\quad r = \dfrac{a}{2} \cot 20° = R \cos 20°$

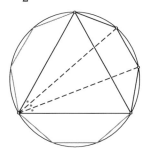

FIGURE 6.2.1

Example 6.2.1. *Drawing a regular nonagon.* Like a regular heptagon, a regular nonagon cannot be drawn with only a straightedge and compass. But if we replace the straightedge with the carpenter's square from Section 3.6 we can easily inscribe a regular nonagon in a circle. Simply inscribe an equilateral triangle in the circle and use the carpenter's square to trisect one of its 60° angles. This trisects the opposite arc, from which all nine sides of the nonagon can be drawn. See Figure 6.2.1. □

The same procedure can be followed trisecting an angle of a regular n-gon to draw a regular $3n$-gon.

In the absence of a carpenter's square (or some other angle trisection device) there exist a variety of approximate angle trisections. Here is a simple compass and straightedge method that is quite accurate for small angles [Steinhaus, 1969].

Example 6.2.2. *Approximate angle trisection.* To trisect the angle $\alpha = \angle AOB$ in Figure 6.2.2, first bisect it and then divide the chord of $\alpha/2 = \angle AOC$ into thirds; the radius to 2/3 of the chord approximately trisects α, i.e., $\beta = \angle AOD \cong \alpha/3$.

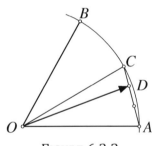

FIGURE 6.2.2

If $O = (0,0)$, $A = (1,0)$, and $C = (\cos(\alpha/2), \sin(\alpha/2))$, then $D = \big((1/3) + (2/3)\cos(\alpha/2), (2/3)\sin(\alpha/2)\big)$ and hence

$$\frac{\alpha}{3} \cong \beta = \arctan\frac{2\sin(\alpha/2)}{1+2\cos(\alpha/2)}.$$

For example, when $\alpha = 60°$,

$$\beta = \arctan\big[1/(1+\sqrt{3})\big] \cong 20.103909°;$$

and when $\alpha = 30°$,

$$\beta = \arctan\big[(\sqrt{6}-\sqrt{2})/(2+\sqrt{6}+\sqrt{2})\big] \cong 10.012765°. \ \square$$

6.2. NONAGONS

In the next example we present an approximate drawing of a regular nonagon by Albrecht Dürer [Hughes, 2012]. Others can be found in [Lenfestey, 1908] and [Running, 1924].

Example 6.2.3. *Albrecht Dürer's 1525 approximation*. In Example 4.5.1 we saw Dürer's approximate construction of a regular heptagon. In the same work he also provided the attractive approximation of a regular nonagon illustrated in Figure 6.2.3. It is almost self-explanatory.

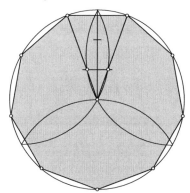

FIGURE 6.2.3

In a circle draw three circular arcs with the same radius from three vertices of an inscribed equilateral triangle (not shown) yielding three "petals." Trisect a radius in one of the petals, and draw a line segment perpendicular to the radius intersecting the two arcs. Projecting the segment to the circle yields one side of the nonagon. Repeating the procedure in the other two petals yields a total of six vertices; the remain three can be found by bisecting arcs of the circle. The side at the top subtends an angle at the center of approximately 39°35′ rather than 40°, so it is somewhat inaccurate. □

A regular nonagon has 27 diagonals, nine short, nine medium, and nine long, of lengths d_1, d_2, and d_3, respectively, as illustrated in Figure 6.2.4. The star polygons inside the regular nonagons in the figure are the three star nonagons, denoted from left to right as $\{9/2\}$, $\{9/3\} = 3\{3\}$, and $\{9/4\}$.

FIGURE 6.2.4

Theorem 6.2.1. *Let a be the side length of a regular nonagon. Then the diagonals have lengths*

(6.1) $\quad d_1 = 2a \cos 20°, d_2 = a(1 + 2 \cos 40°),$ *and*

$$d_3 = 4a \cos 20° \cos 40°.$$

Proof. See Section 1.4.

To approximate the lengths of the diagonals it suffices to approximate $\cos 20°$. In the triple angle formula for the cosine ($\cos 3\theta = 4\cos^3 \theta - 3 \cos \theta$), setting $x = 2 \cos 20°$ yields $x^3 - 3x - 1 = 0$, which has a single positive root. Thus $\cos 20° \cong 0.9397$, $\cos 40° \cong 0.7660$, and (6.1) yields $d_1 \cong 1.8794a$, $d_2 \cong 2.5321a$, and $d_3 \cong 2.8794a$.

Corollary 6.2.2. *Let a, r, R, K, d_1, d_2, and d_3 denote the side length, inradius, circumradius, area, and lengths of the three diagonals in (6.1) of a regular nonagon. Then*

$$d_3 = a + d_1, \sqrt{3}R = d_2, 2ar = Rd_1, \text{ and } K = 3\sqrt{3}d_1d_2/4.$$

Proof. See Challenge 6.2.

Example 6.2.4. *Morrie's law.* The American physicist Richard Feynman (1918-1998) referred to the trigonometric identity

(6.2) $\quad\quad\quad \cos 20° \cos 40° \cos 80° = \dfrac{1}{8}$

as *Morrie's law*, since he heard it from his childhood friend Morrie Jacobs and remembered it for the rest of his life [Beyer et al., 1996]. We prove the identity using a regular nonagon with side length 1 as shown in Figure 6.2.5 [Moreno and García-Caballero, 2015].

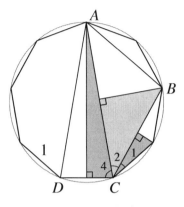

FIGURE 6.2.5

In the figure the angles labeled k have measure $20k°$ for $k = 1, 2,$ and 4. Evaluating the cosine of the marked angle in each gray right triangle yields $\cos 20° = BC/2$, $\cos 40° = AC/2BC$, and $\cos 80° = 1/2AC$, from which the identity follows. See Challenge 6.3 for the sine and tangent versions of Morrie's law. □

Example 6.2.5. *A uniform tiling with nonagons.* While semiregular tilings with nonagons do not exist, a uniform one does. In Figure 6.2.6 we see a portion of a uniform tiling of the plane with regular nonagons and equilateral concave dodecagons (when tiling with concave polygons, corners where only two polygons meet are not counted as vertices).

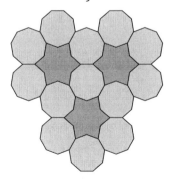

FIGURE 6.2.6

Example 6.2.6. *Nonagons in architecture.* Two beautiful examples of the use of nonagons in architecture are the Bahá'í temples in Figure 6.2.7. The number 9, the final digit in the decimal system, plays an important role in that religion. On the left we see the Lotus Temple in Delhi, India. Its roof has 27 "petals" arranged in sets of three to form the nine sides of the central hall.

FIGURE 6.2.7

On the right we see the Bahá'í House of Worship in Wilmette, Illinois. Like the Lotus Temple, the central hall has nine sides with nine doors, and inside there are nine alcoves and the dome consists of nine sections. □

6.3. Decagons

Decagons are ten-sided polygons, or 10-gons. Below we present some data for a regular decagon with side length a and vertex angle θ, where we express $\pi/10$ radians as $18°$.

vertex angle	$\theta = 4\pi/5 = 144°$
side length	$a = 2r \tan 18° = 2R \sin 18°$
semiperimeter	$s = 5a$
area	$K = \frac{5}{2}a^2 \cot 18° = 10r^2 \tan 18°$
	$= 5R^2 \sin 36° = 5ar$
circumradius	$R = \frac{a}{2} \csc 18°$

6.3. DECAGONS

inradius $\qquad r = \dfrac{a}{2}\cot 18° = R\cos 18°$

From Figure 2.2.1b, Example 2.2.1, and trigonometric identities we have $\sin 18° = \cos 72° = 1/2\varphi = (\sqrt{5}-1)/4$, $\cos 18° = \sin 72° = \sqrt{2+\varphi}/2 = \sqrt{10+2\sqrt{5}}/4$, and $\tan 18° = 1/\sqrt{4\varphi+3} = 1/\sqrt{5+2\sqrt{5}}$. Here are some of the above data, expressed in terms of the golden ratio φ:

side length $\qquad a = R/\varphi$

area $\qquad K = \dfrac{5}{2}a^2\sqrt{4\varphi+3} = 10r^2/\sqrt{4\varphi+3}$
$\qquad\qquad = 5R^2\sqrt{3-\varphi}/2$

circumradius $\ R = a\varphi$

inradius $\qquad r = R\sqrt{2+\varphi}/2$

Since 10 is a product of a power of 2 and the Fermat prime 5, a regular decagon is constructible with straightedge and compass. As noted in Example 2.3.3, Hirano's construction a regular pentagon also constructs a regular decagon. In the following example we present another simple construction.

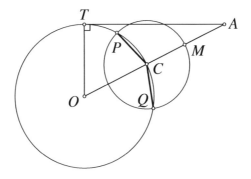

FIGURE 6.3.1

Example 6.3.1. *Another construction of a regular decagon.* In Figure 6.3.1 we have a circle with center O, radius OT, and a line segment $AT = 2OT$ tangent to the circle at T. Draw AO cutting the circle at C, and let M be the midpoint of AC. Draw a circle with center C and radius CM, intersecting the original circle at P and Q. Then line segments CP and CQ are adjacent sides of a regular decagon inscribed in the given circle.

To verify the claim, let $OT = 1$. Then $AT = 2$, $AO = \sqrt{5}$, $AC = \sqrt{5} - 1$, and hence $CM = CP = CQ = (\sqrt{5} - 1)/2 = 1/\varphi$, the side length of a regular decagon inscribed in a circle of radius 1. $\square$

A regular decagon has 35 diagonals of four different lengths, as illustrated in Figure 6.3.2 (only half of the decagon is shown). Note that the five d_1 diagonals are the sides of a regular pentagon whose diagonals have length d_3.

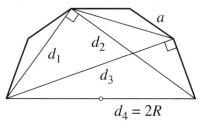

FIGURE 6.3.2

Theorem 6.3.1. *If a, R, d_1, d_2, d_3, and d_4 denote the side length, circumradius, and the lengths of the diagonals of a regular decagon, as illustrated in Figure 6.3.2, then*

$$d_1 = R\sqrt{3 - \varphi}, d_2 = R\varphi, d_3 = R\sqrt{2 + \varphi}, \text{ and } d_4 = 2R.$$

Proof. Since d_1 is the side of a regular pentagon with the same circumradius, we have $d_1 = R\sqrt{3 - \varphi}$. From (1.6) we have $d_2 = a(1 + 2\cos 36°) = a(1 + \varphi) = a\varphi^2 = R\varphi$. Diagonal d_3 satisfies $d_3 = \varphi d_1$, hence $d_3 = R\varphi\sqrt{3 - \varphi} = R\sqrt{2 + \varphi}$, and d_4 is a diameter of the circumcircle. ∎

Corollary 6.3.2. *Under the hypotheses of Theorem 6.3.1 we have*

$$Rd_1 = ad_3, \quad d_1 d_4 = 4ar, \quad \text{and} \quad ad_1 d_2 d_3 = \sqrt{5} R^4.$$

Proof. See Challenge 6.9.

Note that the ratio of the circumradius R to the side length a of a regular decagon is the golden ratio φ. See Challenge 6.6 for a relationship between a and R and the diagonal d_2.

Example 6.3.2. *The inequality $\pi\varphi > 5$.* The perimeter of a polygon inscribed in a circle is less than the circumference of that circle. For a regular decagon with side length $a = R/\varphi$, the perimeter is $10\,R/\varphi$ and the circumference of the circumcircle is $2\pi R$, yielding the inequality $10/\varphi < 2\pi$, or equivalently $\pi\varphi > 5$. □

Example 6.3.3. *The pentagon-hexagon-decagon identity.* This remarkable identity (Proposition XIII.10 in Euclid's *Elements*) states that if a regular pentagon, a regular hexagon, and a regular decagon are inscribed in congruent circles, then their side lengths form a right triangle. See Figure 6.3.3a.

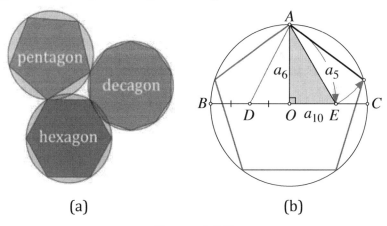

FIGURE 6.3.3

If we let each circle have radius 1 and denote the length of the side of an n-gon by a_n, then $a_{10} = 1/\varphi$, $a_6 = 1$, and $a_5 = \sqrt{3 - \varphi}$. Hence

$$a_{10}^2 + a_6^2 = \frac{1}{\varphi^2} + 1 = 2 - \frac{1}{\varphi} = 3 - \varphi = a_5^2. \quad \Box$$

The right triangle with sides a_5, a_6, and a_{10} also appears in Ptolemy's construction of a regular pentagon in a unit circle from Example 2.3.1. See Figure 6.3.3b.

Example 6.3.4. *Regular decagons in a Penrose aperiodic tiling.* The tilings we have encountered in previous chapters are periodic, in the sense that a set of tiles in a specified region will tile the plane by translation alone. *Aperiodic* tilings are non-periodic tilings without arbitrarily large periodic regions. Among the best known aperiodic tilings are those discovered by the British mathematician Roger Penrose in the 1970s. One example uses two types of rhombi, one with 36° and 144° angles and one with 72° and 108° angles. Both can be found within a regular pentagon, as illustrated on the left in Figure 6.3.4.

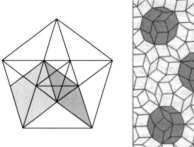
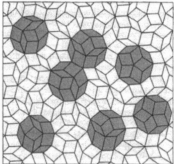

Figure 6.3.4

Of interest is the fact that throughout such an aperiodic tiling regular decagons occur tiled with several patterns of rhombi, as illustrated on the right in Figure 6.3.4. However, all the decagons have the same orientation, i.e., sides of any decagon are parallel to the corresponding sides of every other decagon. □

Example 6.3.5. *Decagons in space.* In Figure 6.3.5 we see a *truncated dodecahedron*, a solid with 12 regular decagonal faces and 20 equilateral triangular faces. This is the fifth truncated Platonic solid, the first four appeared in Sections 3.2 and 5.5.

FIGURE 6.3.5

There are many other solids with regular decagons as faces, e.g., the decagonal prism, antiprism, pyramid, etc.. □

Example 6.3.6. *A decagonal church.* On the left in Figure 6.3.6 we see views of the east elevation and the front door of the Old Zion Baptist Church in Albemarle County, Virginia. It has served the Black community in the county for over one hundred years.

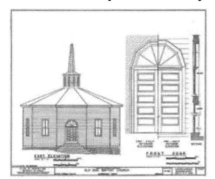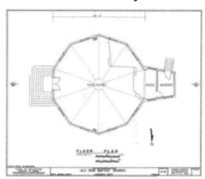

FIGURE 6.3.6

On the right we see the distinctive decagonal floor plan of the sanctuary. Also note that the shape of the fanlight above the front door is one-half of a decagon. □

6.4. Hendecagons

A *hendecagon* (or *undecagon*) is an 11-sided polygon. For data on vertex angle, side length, area, etc., of a regular hendecagon, see Section 1.3. Regular hendecagons cannot be drawn with the Euclidean tools, nor can they be drawn with an angle trisection device. We consider some approximate methods, which have a long history predating the use of computer graphics software.

Consider inscribing a regular hendecagon in a unit circle. Its side length should be $a = 2\sin(\pi/11) \cong 0.563465$. Here are three historic approximate constructions of the side length a of a regular hendecagon with circumradius 1.

Example 6.4.1. *Ancient Greece*. An ancient Greek approximation is $a \cong 14/25 = 0.56$ [Heath, 1981] for a relative error of about 0.615%. □

Example 6.4.2. *Dürer's approximation*. Albrecht Dürer, in his 1525 book *Underweysung der Messung mit dem Zirckel und Richtscheyt* (Instructions on Measurement with Compass and Ruler) suggests the following [Hughes, 2012]: *"To construct an eleven-sided figure by means of a compass, I take a quarter of a circle's diameter, extend it by one-eighth of its length and use this for the construction of the eleven-sided figure."* This yields $a \cong 0.5(1.125) = 0.5625$, for a relative error of about 0.1713%. □

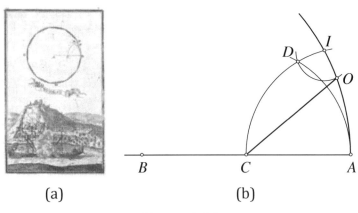

(a)　　　　　　　　　(b)

FIGURE 6.4.1

Example 6.4.3. *Drummond's approximation*. A third approximate method Is described in [Drummond, 1800] as follows: *"Draw the radius AB, bisect it in C—with an opening of the compasses equal to half the radius, upon A and C as centres describe the arcs CDI* *and AD—with the distance ID upon I describe the arc DO and draw the line CO, which will be the extent of one side of an endecagon sufficiently exact for practice."* The figure for Drummond's method is illustrated in the 1698 engraving by Anton Ernst Burkhard von Birckenstein in Figure 6.4.1a. The relevant portion of the unit circle is shown in Figure 6.4.1b. The construction yields $a \cong 0.563692$ for a relative error of about 0.04027%. □

The base of the Statue of Liberty

The Statue of Liberty on Liberty Island in New York harbor sits on the remains of Fort Wood, a former Army base, as seen in Figure 6.4.2a. Some sources claim that the base's fortifications are in the shape of an eleven-pointed star. But the actual shape is an irregular 23-gon with 11 acute-angled vertices and 12 reflex-angled ones, as seen in the plan in Figure 6.4.2b.

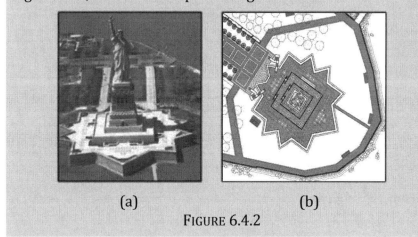

(a) (b)

FIGURE 6.4.2

6.5. Dodecagons

Dodecagons are 12-sided polygons. To draw a regular dodecagon, first inscribe a regular hexagon in a circle to locate six of the 12 vertices, and then bisect the resulting arcs to locate the remaining six vertices. The data below for a regular dodecagon with side length a and vertex angle θ is expressed writing $\pi/12$ radians as 15°.

156   6. MANY-SIDED POLYGONS

vertex angle	$\theta = 5\pi/6 = 150°$
side length	$a = 2r \tan 15° = 2R \sin 15°$
semiperimeter	$s = 6a$
area	$K = 3a^2 \cot 15° = 12r^2 \tan 15° = 3R^2 = 6ar$
circumradius	$R = \frac{a}{2} \csc 15°$
inradius	$r = \frac{a}{2} \cot 15° = R \cos 15°$

To make use of the above data, we first find side lengths of a right triangle with 15° and 75° acute angles using Figure 6.5.1.

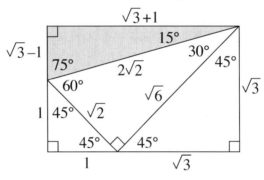

FIGURE 6.5.1

Evaluating functions of 15° in the gray triangle and simplifying yields

$$\tan 15° = \frac{\sqrt{3}-1}{\sqrt{3}+1} = 2 - \sqrt{3}, \quad \sin 15° = \frac{\sqrt{3}-1}{2\sqrt{2}} = \frac{\sqrt{6}-\sqrt{2}}{4}, \quad \text{and}$$

$$\cos 15° = \frac{\sqrt{3}+1}{2\sqrt{2}} = \frac{\sqrt{6}+\sqrt{2}}{4}.$$

Hence we have

side length	$a = 2r(2 - \sqrt{3}) = (\sqrt{6} - \sqrt{2})R/2$
area	$K = 3a^2(2 + \sqrt{3}) = 12r^2(2 - \sqrt{3}) = 3R^2$
circumradius	$R = a(\sqrt{6} + \sqrt{2})/2$
inradius	$r = a(2 + \sqrt{3})/2 = R(\sqrt{6} + \sqrt{2})/4$

Note: Alternate expressions for $\sin 15°$ and $\cos 15°$ are $\sin 15° = \sqrt{2-\sqrt{3}}/2$ and $\cos 15° = \sqrt{2+\sqrt{3}}/2$, which follow from $\sin^2 15° = \left((\sqrt{3}-1)/2\sqrt{2}\right)^2 = (2-\sqrt{3})/4$ and similarly for $\cos^2 15°$. We use these expressions in Section 6.7.

British dodecagonal coins

Prior to decimalization of coinage in Great Britain in 1971, the pound sterling was divided into twenty schillings, and the schilling into twelve pence. A common coin of the time was the *threepence*. From 1937 to 1967 the coin was made from brass in the shape of a regular dodecagon. In Figure 6.5.9 we see the obverse and reverse of the threepence from its final year in general circulation.

FIGURE 6.5.9

In 2017 the Royal Mint produced a similar bimetallic (nickel and brass) dodecagonal one pound coin.

A regular dodecagon has 54 diagonals of five different lengths, as illustrated in Figure 6.5.2.

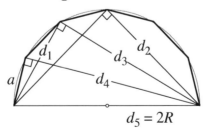

FIGURE 6.5.2

Since the longest diagonal is a diameter of the circumcircle and the other diagonals are sides of a regular hexagon, square, or equilateral triangle with the same circumcircle, the lengths are readily computed:

$$d_1 = R, \quad d_2 = R\sqrt{2}, \quad d_3 = R\sqrt{3}, \quad d_4 = 2r = R(\sqrt{6}+\sqrt{2})/2, \text{ and}$$
$$d_5 = 2R.$$

Example 6.5.1. *The area of a regular dodecagon*. The area formula $K = 3R^2$ can be illustrated visually as shown in Figure 6.5.3 where we partition a regular dodecagon with circumradius R into 12 triangles which can be reassembled to form three congruent squares each with area R^2. □

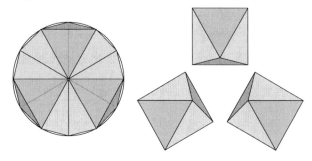

FIGURE 6.5.3

Example 6.5.2. *A dissection proof of the area formula for a regular dodecagon*. In Figure 6.5.4 we construct a square by drawing lines through eight vertices of a regular dodecagon. Since d_3 is the side length of the square, the area of the square is the same as the area of the dodecagon, i.e., $d_3^2 = (\sqrt{3}R)^2 = 3R^2 = K$. Note that the area of each light gray triangle is $a^2/4$ while the area of each dark gray triangle is $(1/2)a^2 \sin 150° = a^2/4$.

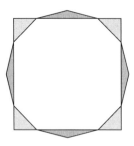

FIGURE 6.5.4

6.5. DODECAGONS

The fact that the area of the dodecagon equals d_3^2 leads to the following six-piece dissection and reassembly of the pieces to form the square as shown in Figure 6.5.5. [Lindgren, 1951]. □

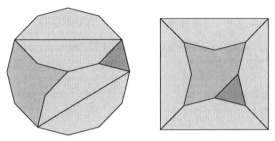

FIGURE 6.5.5

The Wallace-Bolyai-Gerwien theorem

The dissection in Figure 6.5.5 is an example of a result known as the *Wallace-Bolyai-Gerwien theorem*, named for William Wallace (1768–1843), Farkas Bolyai (1775–1856), and Paul Gerwien (1799–1858), who found proofs independently and published them in 1831, 1833, and 1835, respectively. The theorem states that given two polygons in the plane with the same area, each one can be dissected into finitely many smaller polygons that can then be reassembled to form the other. The polygons need not be convex nor have the same number of sides. In Figure 6.5.6 we have dissections of a regular octagon and a square. See [Frederickson, 1997] for illustrations of many other such dissections.

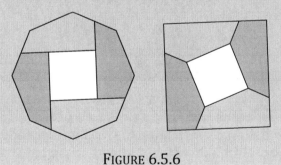

FIGURE 6.5.6

Example 6.5.3. *Dodecagons in semiregular tilings.* In Sections 1.7, 3.4, and 5.2 we have seen some of the eight semiregular tilings of the plane. Two of those tilings employ regular dodecagons; portions of those tilings are illustrated in Figure 6.5.7. □

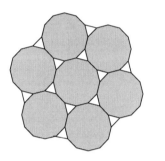 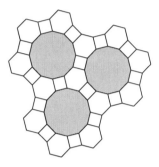

FIGURE 6.5.7

Example 6.5.4. *A dodecagonal tower.* Towers in the shape of a polygonal prism are fairly common. A beautiful example of a dodecagonal tower is the *Torre del Oro* ("tower of gold") along the banks of the Guadalquivir River in Seville, Spain, as seen in Figure 6.5.8.

FIGURE 6.5.8

The tower is 36 meters tall and was constructed in the early 13th century, although the middle dodecagonal section was added in the 14th century and the uppermost circular cylindrical portion was added in the 18th century. □

6.6. Gauss and heptadecagons

A *heptadecagon* is a 17-sided polygon. It is of interest since $17 = 2^{2^2} + 1$ is a Fermat prime, and hence a regular heptadecagon is constructible with straightedge and compass, a fact proven by Carl Friedrich Gauss in 1796 when he was 19 years old. The proof is based on the fact that the constructability of an *n*-gon is equivalent to being able to express trigonometric functions of the angle at the center subtended by a side of the *n*-gon in terms of arithmetic functions and square roots. In his book *Disquisitiones Arithmeticae* Gauss showed that

$$16 \cos \frac{2\pi}{17} = -1 + \sqrt{17} + \sqrt{34 - 2\sqrt{17}} + 2\sqrt{17 + 3\sqrt{17} - \sqrt{34 - 2\sqrt{17}} - 2\sqrt{34 + 2\sqrt{17}}}.$$

Gauss did not provide a construction of the 17-gon, the first to do so was Johannes Erchinger in 1800. We present a later construction by the English mathematician Herbert William Richmond (1863-1948) [Richmond, 1893].

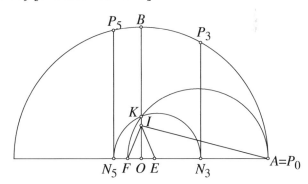

FIGURE 6.6.1

Draw a circle with center O and radius OA as shown in Figure 6.6.1 (only the upper semicircle is illustrated), and draw OB perpendicular to OA at O. Locate I on OB so that OI equals $1/4$ of OB, and draw AI. Locate E on OA so that $\angle OIE$ equals $1/4$ of

$\angle OIA$ and locate F on the diameter so that $\angle EIF = 45°$. Draw a circle with diameter AF intersecting OB at K. Draw a circle with radius EK centered at E and intersecting the diameter at N_3 and N_5. Draw lines perpendicular to the diameter at N_3 and N_5 intersecting the semicircle at P_3 and P_5. With $A = P_0$ the arcs P_0P_3 and P_0P_5 are 3/17 and 5/17 of the circumference of the circle, from which the heptadecagon can be drawn.

In Figure 6.6.2 we have a 1977 postage stamp from the German Democratic Republic with a portrait of Gauss, a compass, a straightedge, and a 17-gon.

FIGURE 6.6.2

6.7. Archimedes and 24-, 48-, and 96-gons

Polygons with 96 sides are important in the history of mathematics, as they were employed by Archimedes of Syracuse (circa 287-212 BCE) in his book *Measurement of a Circle* to produce his remarkable inequality $3\frac{10}{71} < \pi < 3\frac{1}{7}$. If we let p_n and P_n denote respectively the perimeters of n-gons inscribed in and circumscribed about a unit circle, Archimedes used the recursions

$$P_{2n} = \frac{2p_nP_n}{p_n+P_n} \quad \text{and} \quad p_{2n} = \sqrt{p_nP_{2n}}$$

(see Theorem 1.3.1) along with $p_6 = 6$ and $P_6 = 4\sqrt{3}$ to compute $p_{12}, P_{12}, p_{24}, P_{24}, p_{48}, P_{48},$ and p_{96}, P_{96} [Boyer, 1968].

Archimedes and the Fields Medal

In Figure 6.7.1 we have an image of the Fields Medal with the head of Archimedes. The Medal, perhaps the most coveted award for mathematicians, is a prize awarded every four years by the International Mathematical Union to two to four mathematicians under the age of 40 for outstanding mathematical achievement.

FIGURE 6.7.1

In designing the medal, the artist wrote that he wanted to give "the mathematical world a version of Archimedes which is not decrepit, bald-headed, and myopic, but which has the fine presence and assured bearing of the man who defied the power of Rome" [Tropp. 1976].

We present a procedure equivalent to Archimedes' but with simpler recursions; computing side lengths a_n and A_n of the inscribed and circumscribed n-gons, respectively, from which the perimeters readily follow. The desired approximations to π are the semiperimeters $48a_{96}$ and $48A_{96}$, i.e., $48a_{96} < \pi < 48A_{96}$.

The recursions are based on the half-angle formulas for the sine and cosine. For convenience let $s_n = \sin(\pi/n)$, $c_n = \cos(\pi/n)$, and $t_n = \tan(\pi/n) = s_n/c_n$. Then

$$s_{2n} = \sqrt{\frac{1-c_n}{2}} = \frac{1}{2}\sqrt{2-2c_n} \quad \text{and} \quad c_{2n} = \sqrt{\frac{1+c_n}{2}} = \frac{1}{2}\sqrt{2+2c_n}.$$

Since $c_6 = \sqrt{3}/2$ we have $s_{12} = \frac{1}{2}\sqrt{2 - \sqrt{3}}$ and $c_{12} = \frac{1}{2}\sqrt{2 + \sqrt{3}}$. Continuing in this fashion yields

$$s_{24} = \frac{1}{2}\sqrt{2 - \sqrt{2 + \sqrt{3}}} \text{ and } c_{24} = \frac{1}{2}\sqrt{2 + \sqrt{2 + \sqrt{3}}},$$

$$s_{48} = \frac{1}{2}\sqrt{2 - \sqrt{2 + \sqrt{2 + \sqrt{3}}}} \text{ and } c_{48} = \frac{1}{2}\sqrt{2 + \sqrt{2 + \sqrt{2 + \sqrt{3}}}},$$

$$s_{96} = \frac{1}{2}\sqrt{2 - \sqrt{2 + \sqrt{2 + \sqrt{2 + \sqrt{3}}}}} \text{ and}$$

$$c_{96} = \frac{1}{2}\sqrt{2 + \sqrt{2 + \sqrt{2 + \sqrt{2 + \sqrt{3}}}}}.$$

Since the polygons are inscribed in and circumscribed about a unit circle, we have $a_{96} = 2s_{96}$ and $A_{96} = 2t_{96} = 2s_{96}/c_{96}$, so that the desired semiperimeters of the inscribed and circumscribed 96-gons are

$$48a_{96} \cong 3.1410319509 \text{ and } 48A_{96} \cong 3.1427145997.$$

Since $48a_{96} = 96\sin(\pi/96)$ and $48A_{96} = 96\tan(\pi/96)$, the inequality $48a_{96} < \pi < 48A_{96}$ is equivalent to $\sin\theta < \theta < \tan\theta$ for $\theta = \pi/96$.

Using approximations to $\sqrt{3}$ Archimedes was able to show that

$$3\tfrac{10}{71} = 3\frac{284\tfrac{1}{4}}{2018\tfrac{7}{40}} < 3\frac{284\tfrac{1}{4}}{2017\tfrac{7}{40}} < 48a_{96} < \pi < 48A_{96} < 3\frac{667\tfrac{1}{2}}{4673\tfrac{1}{2}} < 3\frac{667\tfrac{1}{2}}{4672\tfrac{1}{2}} = 3\tfrac{1}{7}$$

to conclude $3\tfrac{10}{71} < \pi < 3\tfrac{1}{7}$ [Struik, 1967].

Example 6.7.1. *In Archimedes' footsteps.* In the centuries after Archimedes, many mathematicians used perimeters of a variety

6.7. ARCHIMEDES AND 24-, 48-, AND 96-GONS

of regular many-sided polygons to establish better approximations of π. Here is a brief summary. See [Schepler, 1950] and [Lam and Ang, 1986] for details and references.

- 263 CE. Liu Hui (China). $\pi \cong 3.14$ by inscribing a regular 192-gon, and $\pi \cong 3.1416$ by inscribing a regular 3072-gon (all the polygons mentioned below are regular).

- 480 CE. Zu Chongzhi (China). $3.1415926 < \pi < 3.1415927$ by inscribing 12288- and 24576-gons. Zu also observed that $\pi \cong 355/113$.

- 500 CE. Aryabhata (India). $\pi \cong 3.1416$, perhaps by inscribing a 384-gon.

- 1579. François Viète (France). $\pi \cong 3.141592653$, i.e., correct to 9 places, using a $3 \cdot 2^{17}$-gon. (Note that $3 \cdot 2^{17} = 393,216$.)

- 1593. Adriaan van Rooman (Netherlands). π correct to 15 places, circumscribing a 2^{30}-gon ($2^{30} = 1,073,741,824$.)

- 1596. Ludolph van Cuelen (Germany). π correct to 20 places, inscribing and circumscribing $60 \cdot 2^{33}$-gons. ($60 \cdot 2^{33} = 515,396,075,520$.)

- 1620. Ludolph van Cuelen again. π correct to 35 places, using a 2^{62}-gon. ($2^{62} = 4,611,686,018,427,387,904$.)

Archimedes' method of approximating π was superseded in the 17th century when the Dutch mathematicians Christiaan Huygens and Willebrord Snellius found ways to approximate lengths of circular arcs that were much more efficient than using regular polygons. □

In Figure 6.7.2 we see a page from a 16th century Ming dynasty edition of Liu Hui's *Jiuzhang suanshu* (*Nine Chapters on the Mathematical Art*) featuring an image describing his approximation of π with a regular dodecagon.

FIGURE 6.7.2

6.8. The 257-gons and the 65537-gons

The regular 257-gon is of interest since $257 = 2^{2^3} + 1$ is a Fermat prime, hence a 257-gon is constructible with straightedge and compass. The same is true for the regular 65537-gon, since $65537 = 2^{2^4} + 1$ is also a Fermat prime.

Although Gauss proved in 1801 that the regular 257-gon was constructible with straightedge and compass, he did not provide such a construction. The German mathematician Friedrich Julius Richelot (1808–1875) published a 194-page construction of the regular 257-gon in a four-part paper in *Crelle's Journal* in 1832 [DeTemple, 1999]. With the aid of a computer a modern construction using complex numbers is possible, see [Gottlieb, 1999] for details.

Constructing a 65537-gon is another matter entirely. For all practical purposes it is indistinguishable from a circle. Concerning Gauss's request for a regular 17-gon on his tombstone [Boyer, 1968; Eves, 1983], the American mathematician Eric Temple Bell (1883–1960) opined [Bell, 1951]:

> "If he had requested the like for the regular polygon of 65,537 sides, his executors might have had to duplicate or surpass the Great Pyramid. Of course this is an exaggeration. But nobody yet has

been so pertinaciously stupid to actually carry out the construction for 65,537. Happily a Euclidean construction for 4,294,967,297 is impossible."

Bell appears to have been unaware that the German mathematician Johann Gustav Hermes (1846–1912) worked for ten years on the construction of the regular 65537-gon. His manuscript, with over 200 pages, is now in a trunk in the attic at the Mathematical Institute in Göttingen, perhaps never having been read [DeTemple, 1999].

6.9. Miscellaneous many-sided polygons

In this section we present a collection of examples employing a variety of *n*-gons with applications. The examples are in order of increasing *n*.

Example 6.9.1. *The 20-80-80 triangle and the 18-gon.* The so-called 20-80-80 (or 80-80-20) triangle problem is about a century old, perhaps having first appeared on page 173 in the October 1922 issue of the *Mathematical Gazette*. It has many variants, and has been solved in many ways. The angles in the problem (and in the many variants) are all multiples of $10°$, and the solution we present [Prasolov, 2000] employs a regular *octadecagon*, or 18-gon, since many angles between its sides and diagonals are multiples of $10°$.

Here is the problem. In isosceles triangle ABC with base AC and $\angle ABC = 20°$ choose points D and E on sides BC and AB respectively so that $\angle DAC = 60°$ and $\angle ECA = 50°$, as shown in Figure 6.9.1a. Prove that $\angle ADE = 30°$.

In Figure 6.9.1b we have the lower portion of the triangle imbedded in a regular 18-gon (only its circumcircle and vertices are shown). Extensions of segments AE, CE, and DE are concurrent diagonals of the 18-gon, as shown in the tiling by 18 congruent irregular equilateral pentagons in Figure 6.9.1c (see [Prasolov, 2000] for a formal proof).

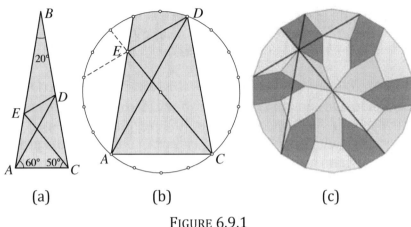

FIGURE 6.9.1

Thus ∠ADE subtends 3/18 of the circumference of the circumcircle of the 18-gon, i.e., ∠ADE = 30°. □

The pentagons in Figure 6.9.1c are Type 1 pentagons from Section 2.6, with angles (clockwise) 60°, 160°, 80°, 100°, and 140°. Thus each pentagon is the union of an equilateral triangle and a rhombus with 80° and 100° angles.

See Challenge 6.15 for another triangle problem related to the 18-gon.

Example 6.9.2. *A surprising property of the regular 26-gon.* Let $P_1 P_2 P_3 \cdots P_{26}$ be a regular 26-gon inscribed in a unit circle with center O. Let X and Y be the reflections of O in the chords $P_1 P_3$ and $P_4 P_8$, respectively, as shown in Figure 6.9.2. Then XY is equal in length to a side of an equilateral triangle inscribed in the circle.

An equilateral triangle inscribed in a unit circle has side length $\sqrt{3}$, thus we must show that $XY = \sqrt{3}$. Let $\alpha = 2\pi/26 = \pi/13$, so that $OX = 2\cos\alpha$, $OY = 2\cos 2\alpha$, and $\angle XOY = 4\alpha$. The law of cosines yields

$$XY^2 = 4\cos^2\alpha + 4\cos^2 2\alpha - 8\cos\alpha \cos 2\alpha \cos 4\alpha.$$

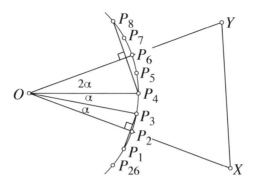

FIGURE 6.9.2

Since $2\cos^2\alpha = 1 + \cos 2\alpha$ and $2\cos^2 2\alpha = 1 + \cos 4\alpha$ we have

$$XY^2 = 4 + 2\cos 2\alpha + 2\cos 4\alpha - 8\cos\alpha \cos 2\alpha \cos 4\alpha.$$

Multiplying both sides by $\sin\alpha$ and noting that $2\sin\alpha\cos 2\alpha = \sin 3\alpha - \sin\alpha$, $2\sin\alpha\cos 4\alpha = \sin 5\alpha - \sin 3\alpha$, and $8\sin\alpha\cos\alpha\cos 2\alpha\cos 4\alpha = \sin 8\alpha$ yields

$$XY^2 \sin\alpha =$$
$$4\sin\alpha + (\sin 3\alpha - \sin\alpha) + (\sin 5\alpha - \sin 3\alpha) - \sin 8\alpha$$
$$= 3\sin\alpha + \sin 5\alpha - \sin 8\alpha.$$

Since $5\alpha + 8\alpha = \pi$ we have $\sin 5\alpha = \sin 8\alpha$, hence $XY^2 = 3$, which completes the proof [Thébault and Blundon, 1958]. □

Example 6.9.3. *Drawing a regular hexacontagon*. A regular *hexacontagon* (60-gon) is constructible with straightedge and compass since $60 = 2^2 \cdot 3 \cdot 5$ and 3 and 5 are Fermat primes. Here is a particularly simple two-circle construction inscribing one side of the 60-gon in a given circle, which has the added benefit of inscribing sides of a regular *pentadecagon* (15-gon) and a regular pentagon [Lord, 2012].

In Figure 6.9.3 AO and BO are perpendicular radii of the given circle, and C is the midpoint of AO. The circle with center C and radius BC passes through D, E, and F, where F (close to B)

satisfies $EF = BC$. Then chord BH is the side of an inscribed regular 60-gon, chord AG is the side of an inscribed regular 15-gon, and chord GH is the side of an inscribed regular pentagon.

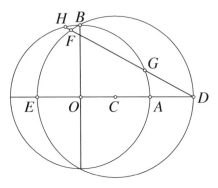

FIGURE 6.9.3

To verify the construction, first note that $\angle ODF = 30°$ since $ED = 2EF$ in right triangle $\triangle DEF$. Applying the law of sines to $\triangle OHD$ yields

$$\sin \angle OHD = \frac{OD \sin 30°}{OH} = \frac{1+\sqrt{5}}{4} = \cos 36° = \sin 54°$$

(recall Example 2.2.1) so that $\angle OHD = 54° = \angle OGH$. It now sfollows that $\angle OGD = 126°$ and $\angle DOH = 180° - (30° + 54°) = 96°$, so that $\angle BOH = 6°$. Then $\angle AOG = 180° - (30° + 126°) = 24°$, and $\angle GOH = 180° - 2 \cdot 54° = 72°$, as claimed. $\square$

6.10. Challenges

6.1 For a regular nonagon show that the medium length diagonal d_2 is given by $d_2 = (\sqrt{3}/2) a \csc 20°$.

6.2 Prove Corollary 6.2.2.

6.3 Prove the following analog of Morrie's law (6.2) for sines:

$$\sin 20° \sin 40° \sin 80° = \frac{\sqrt{3}}{8}.$$

(Hint. Express 4sin 40° sin 80° in terms of sin 20° and use the triple angle formula for the sine.) Note that the tangent version then follows:

$$\tan 20° \tan 40° \tan 80° = \sqrt{3}.$$

6.4 Let a, R, d_1, d_2, and d_3 denote the side length, circumradius, and lengths of the three diagonals of a regular nonagon. Show that $ad_1 d_2 d_3 = 3R^4$.

6.5 Let AB and BC be adjacent sides of a regular nonagon inscribed in a circle with center O, as shown in Figure 6.10.1. Let M be the midpoint of arc AB, N the midpoint of side BC, and P the midpoint of OM. Show that $\angle PNO = 30°$.

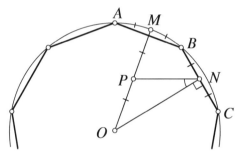

Figure 6.10.1

6.6 Within a regular decagon and its circumcircle, draw the diagonals d_2 that join every third vertex, as shown in Figure 6.10.2

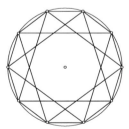

Figure 6.10.2

Show that the length d_2 of one of these diagonals is equal to the sum of the side length a and the circumradius R.

6.7 Let x be a positive number not equal to 1 such that x, x^2, and x^4 are in arithmetic progression. Show that x equals the side length of a regular decagon inscribed in a unit circle.

6.8 An equilateral triangle is surrounded by a chain of six congruent circles in such a manner that three of the circles touch the vertices and the other three touch the midpoints of the sides of the triangle, as shown in Figure 6.10.3. Show that the inradius of the triangle is equal to the side length of a regular decagon inscribed in one of the surrounding circles.

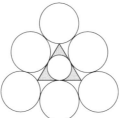

Figure 6.10.3

6.9 Prove Corollary 6.3.2.

6.10 Show that a regular decagon can be partitioned into ten congruent equilateral pentagons. (Hint. See Challenge 2.8.)

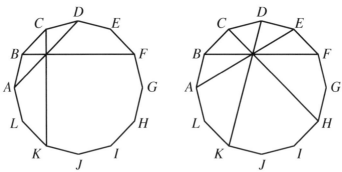

Figure 6.10.4

6.11 Let $ABCDEFGHIJKL$ be the regular dodecagon shown (twice) in Figure 6.10.4. Show that (i) the diagonals AD, BF, and CK are concurrent, and (ii) the diagonals AE, BF, CH, and *DK* are concurrent. (Hint. Dissect the dodecagon into six equilateral triangles, six squares, and one hexagon.)

6.12 Use the square with side length 2 in Figure 6.10.5a to show that the area of a regular dodecagon with circumradius 1 equals 3.

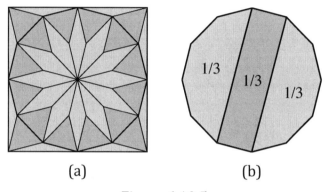

(a)   (b)

Figure 6.10.5

6.13 (i) Show that a pair of parallel d_4 diagonals partition the area of a regular dodecagon into thirds, as shown in Figure 6.10.5b. (ii) Is it possible to partition a regular dodecagon into three congruent octagons?

6.14 Choose points *X* and *Y* on sides *BC* and *CD*, respectively, of square *ABCD* so that $\angle BAX = \angle XAY = \angle YAD = 30°$ as shown in Figure 6.10.6. Show that the area of the dark gray triangle *AXY* is 1/3 the area of the square. (Hint. Consider the upper right quadrants of the dodecagon and square in Figure 6.5.4.)

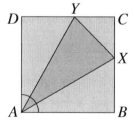

Figure 6.10.6

6.15 In isosceles triangle ABC with base BC and ∠BAC = 80° choose M inside the triangle with ∠MBC = 30° and ∠MCB = 10°, as shown in Figure 6.10.7. Prove that ∠AMC = 70°. (Hint. Let A, B, and C be vertices of a regular 18-gon.)

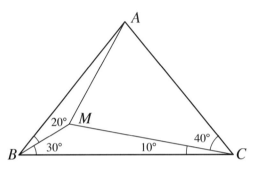

Figure 6.10.7

6.16 Choose any nine vertices of a regular icosagon (a 20-sided polygon). Show that three of those vertices will always form an isosceles triangle. (Hint. Label the vertices A, B, C, D, A, B, C, D, ... around the icosagon. How many of the nine vertices have the same label?)

6.17 A hexagram inscribed in a regular hexagon with side length a has area $a^2\sqrt{3}$, as does an equilateral triangle with side length $2a$. The Wallace-Bolyai-Gerwien theorem guarantees a dissection of the hexagram and rearrangement of the pieces to form the triangle. Find such a dissection. (Hint: A five-piece dissection is possible, one piece of which is the gray hexagon in Figure 6.10.8.)

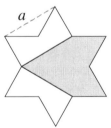

Figure 6.10.8

CHAPTER 7

Miscellaneous Classes of Polygons

From such great heights she glimpses the enormous shapes stamped on the earth, the long polygons made by the borders of farms and rivers and mill towns, littoral masses and city walls, a reflection of the celestial mosaic.

Catherynne M. Valente, *Palimpsest*

7.1. Introduction

In this chapter we discuss a variety of classes of interesting polygons whose members share common properties . We begin with two types defined in the Cartesian plane: *lattice polygons*, polygons whose vertices are lattice points; and *rectilinear polygons*, polygons whose sides are parallel to the coordinate axes. Next we present the *zonogons*, which are generalizations of the parallelogons (e.g., parallelograms and parahexagons) from Section 3.5. In the final two sections of this chapter we study properties of the classes of *cyclic polygons* and *tangential polygons*, classes that contain not only the regular polygons but also many irregular polygons with intriguing properties.

7.2. Lattice polygons

A *lattice point* in the plane is a point with integer coordinates, and a *lattice polygon* is a polygon whose vertices are lattice points. A polygon is *simple* if it has no self-intersections. Of interest in this section we explore areas of a simple lattice polygons.

Example 7.2.1. *Lattice squares*. In Figure 7.2.1 we see seven lattice squares with the indicated areas.

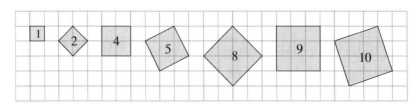

FIGURE 7.2.1

The sequence 1, 2, 4, 5, 8, 9, 10, ... , of areas is sequence A001481 in the *Online Encyclopedia of Integer Sequences* at oeis.org. The elements of the sequence are the positive integers equal to the sum of the squares of two integers (since the side length is the distance between two lattice points). □

Example 7.2.2. *A lattice triangle.* In Figure 7.2.2 we see a lattice triangle with side lengths $a = \sqrt{10}$, $b = \sqrt{13}$, and $c = \sqrt{17}$.

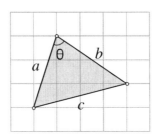

FIGURE 7.2.2

Using Heron's formula, its area K is

$$K = \frac{1}{4}\sqrt{(a+b+c)(-a+b+c)(a-b+c)(a+b-c)},$$

which after some algebra becomes

$$K = \frac{1}{4}\sqrt{(2ab)^2 - (a^2+b^2-c^2)^2} = \frac{1}{4}\sqrt{520-36} = \frac{11}{2}.$$

Alternatively, we can compute the area using the lengths of two sides and the sine of the included angle. Let θ denote the measure of the angle between the sides with lengths a and b. Then $\theta = \arctan(1/3) + \arctan(3/2)$ so that $\tan\theta = 11/3$ and thus $\sin\theta = 11/\sqrt{130}$. Thus $K = (1/2)\sqrt{10}\sqrt{13}\sin\theta = 11/2$. □

However, there is a straightforward method to compute the area of a simple lattice polygon. This gem of elementary geometry was discovered in 1899 by the Austrian mathematician Georg Alexander Pick (1859–1942). Pick was born in Vienna but lived most of his life in Prague. He was captured by the Nazis in 1942 and sent to the Theresienstadt concentration camp where he perished.

Pick's theorem expresses the area of a simple lattice polygon in terms of a linear function of the number of lattice points on the boundary and the number in the interior of the polygon. Today schoolchildren are often introduced to Pick's theorem using *geoboards* (wooden or plastic boards with pegs) and rubber bands to form polygons and calculate areas. Geoboards were introduced and illustrated by Caleb Gattegno in [Gattegno, 1971].

Pick's Theorem 7.2.1. *Let S be a simple lattice polygon with i internal lattice points and b boundary lattice points. Then the area K of S is given by*

$$K = \frac{b}{2} + i - 1.$$

There are a variety of ways to prove Pick's theorem, all nontrivial. The proof in [Varberg, 1985], which we will not reproduce here, is rather direct and intuitive and can be found in [Alsina and Nelsen, 2010]. See [Grünbaum and Shephard, 1993] for references to a number of other proofs.

Observe that for the triangle in Figure 7.2.2 we have $b = 3$ and $i = 5$ so that $K = (3/2) + 5 - 1 = 11/2$.

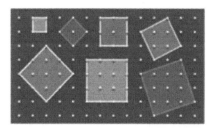

Georg Pick and a geoboard version of Figure 7.2.1.

Example 7.2.3. *Approximations to π with simple lattice polygons.* With Pick's theorem we illustrate two classical approximations to π, the Babylonian value 25/8 (see Challenge 3.15) and the Archimedean value 22/7. In each case we approximate a quarter circle with a lattice polygon.

In Figure 7.2.3a we have a quarter circle with radius 8 and a lattice hexagon. Counting boundary and interior points yields $b = 20$ and $i = 41$ so that $K = 50$. Thus the approximate value $\hat{\pi}$ of π satisfies $8^2\hat{\pi}/4 = 50$, so that $\hat{\pi} = 25/8$.

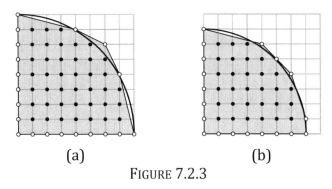

(a)   (b)

FIGURE 7.2.3

In Figure 7.2.3b we have a quarter circle with radius 7 and another lattice hexagon. Counting boundary and interior points yields $b = 19$ and $i = 30$ so that $K = 77/2$. Thus the approximate value $\hat{\pi}$ of π satisfies $7^2\hat{\pi}/4 = 77/2$, so that $\hat{\pi} = 22/7$. □

Example 7.2.4. *Relationships between the numbers of interior and boundary points.* What values for b and i are possible for simple convex lattice polygons? First note that for a given value of b, i can be arbitrarily large. For example, a lattice triangle with $b = 3$ (e.g., the one in Figure 7.2.2) can contain arbitrarily many interior points. In the other direction, when $i = 0$ there can be arbitrarily many boundary points, (e.g., the one in Figure 7.2.4).

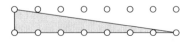

FIGURE 7.2.4

However, when $i = 1$ the number b of boundary points must lie between 3 and 9 [Scott, 1976], and examples of such convex lattice triangles, squares, pentagons, and hexagons are shown in Figure 7.2.5.

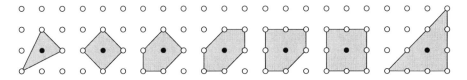

FIGURE 7.2.5

When $i \geq 2$ the number b of boundary points lies between 3 and $2i + 6$ [Scott, 1976]. In Figure 7.2.6 we illustrate the bounds 3 and 14 for the case $i = 4$.

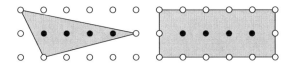

FIGURE 7.2.6

Similar inequalities exist relating the area K to b or i; see [Haase and Schicho, 2009]. □

Example 7.2.5. *Lattice pentagons with minimum area.* If S is a convex simple lattice pentagon, then its area is at least $5/2$ (proving this is Problem A3 from the 1990 William Lowell Putnam Mathematical Competition).

Consider a convex simple lattice pentagon *ABCDE* of minimum area. By Pick's theorem the area of a simple lattice polygon must be half an integer, so a lattice pentagon with minimum exists. If the interior of side *AB* contains a lattice point *F*, then *AFCDE* is a simple lattice pentagon with smaller area, contradicting the choice of *ABCDE*. Applying this argument to each side we may assume that all boundary lattice points of *ABCDE* are vertices.

Separate the five vertices into four classes according to the parity of their coordinates (i.e., both even, both odd, first even second odd, and first odd second even). One class must contain at least two vertices by the pigeonhole principle. The midpoint M between two such vertices has integer coordinates. By the argument in the previous paragraph these two vertices cannot be a side of the pentagon, so M is an interior point since the pentagon is convex.

Connecting M to the vertices partitions the pentagon into 5 triangles, each with area at least $1/2$, so the whole pentagon has area at least $5/2$. The bound cannot be improved, since the pentagon with vertices $(0,0)$, $(1,0)$, $(2,1)$, $(1,2)$ and $(0,1)$ is convex and has area $5/2$ by Pick's theorem. See Figure 7.2.7a [Kedlaya et al., 2012]. □

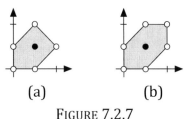

(a) (b)

FIGURE 7.2.7

The corresponding problem for a convex simple lattice hexagon yields a minimum area of 3, attained by the lattice parahexagon in Figure 7.2.7b. See [Arkinstall, 1980] for details.

Example 7.2.6. *Lattice heptagons and octagons with minimum area.* If S is a convex simple lattice heptagon, then its area is at least $13/2$, and if S is a convex simple lattice octagon, then its area is at least 7 [Rabinowitz, 1990]. In these cases the lattice polygons have four interior lattice points. The bounds cannot be improved, as shown by the lattice polygons in Figure 7.2.8. □

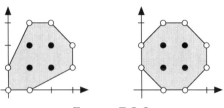

FIGURE 7.2.8

7.3. Rectilinear polygons

A *rectilinear* (or *orthogonal*) *polygon* is one in which every internal vertex angles measure either 90° or 270° (by a slight abuse of notation, we refer to these as 90° vertices and 270° vertices). Rectilinear polygons are those that, when suitably placed on a coordinate plane, have sides that are parallel to the axes. Consequently all rectangles and all polyominoes (see Section 3.7) are rectilinear. A *simple* (or *hole-free*) rectilinear polygon is one that has no holes.

Every simple rectilinear polygon has an even number of sides since in a traversal of its perimeter, each horizontal side is followed by a vertical side and conversely. Furthermore, if a is the number of 90° vertices and b is the number of 270° vertices in a rectilinear $2n$-gon, then $a = b + 4$, i.e., $a = n + 2$ and $b = n - 2$. See Challenge 7.4.

Example 7.3.1. *Rectilinear hexagons and octagons.* A simple rectilinear hexagon has five 90° vertices and a single 270° vertex, i.e., it is an *L-hexagon* as seen in Figure 3.2.1. A representative of this class of hexagons is the *L-polyomino* illustrated in Figure 7.3.1a.

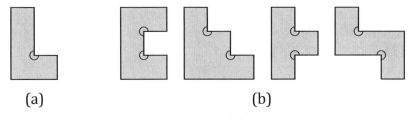

(a)   (b)

FIGURE 7.3.1

A simple rectilinear octagon has six 90° vertices and two 270° vertices, and has four possible shapes, depending on the number of 90° vertices between the two 270° vertices. That number can be 0, 1, 2, or 3, as shown in Figure 7.3.1b with four polyominoes. Other members of each class are *vertex-equivalent* (vertices in the same order with the same size angles) to the polyomino illustrated, but with different lengths for some sides. □

A rectilinear hexagonal residence

In Figure 7.3.2 we see an illustration of a 17th century Japanese estate, consisting of three rectangular units with various buildings within. The resulting outline of the estate is a rectilinear hexagon. Even today many architectural entities are shaped like rectilinear polygons.

FIGURE 7.3.2

Example 7.3.2. *Golygons*. A *golygon* (also called a *serial isogon of 90 degrees*) is a simple rectilinear polygon whose side lengths in order are consecutive integers. The smallest is the octagon In Figure 7.3.3a whose side lengths in order are 1 through 8. Note that this golygon is vertex-equivalent to the rightmost polyomino in Figure 7.3.1b.

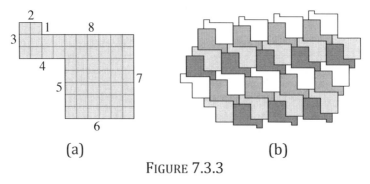

(a)　　　　　　　　　　(b)

FIGURE 7.3.3

See [Sallows et al., 1991] for a discussion of golygon properties, e.g., every rectilinear golygon is an $8n$-gon, and the only golygon known to tile the plane is the one above, as shown in Figure 7.3.3b.

However, not all polygons whose side lengths in order are consecutive integers are rectilinear. The nonagon in Figure 7.3.4 is a *serial isogon of 60 degrees* based on the triangular lattice, whose side lengths in order are 1 through 9. For more details, see [Sallows, 1992]. □

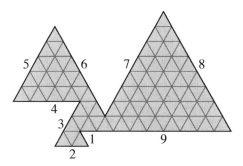

FIGURE 7.3.4

7.4. Zonogons

A *zonogon* is a convex polygon with an even number of sides and the property that opposite sides are parallel and have equal length. Consequently a zonogon is also known as a *centrally symmetric convex polygon*. Examples include regular $2n$-gons, parallelograms (including squares, rectangles, and rhombi), convex parahexagons (see Section 3.5), and the irregular octagon in Theorem 5.3.1.

Example 7.4.1. *Dissecting a regular hexagon into zonogons.* In Figure 7.4.1a we see a regular hexagon dissected into three rhombi. This is an example of an *irreducible* dissection, one such that no non-singleton proper subset of the polygons in the dissection forms a zonogon.

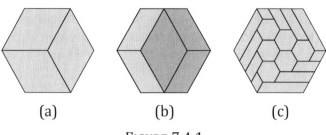

(a)   (b)   (c)

FIGURE 7.4.1

In Figure 7.4.1b we see a dissection into five parts, all zonogons, that is not irreducible, in that the three polygons shaded dark gray form a zonogon. So the question arises: how many different irreducible dissections of the regular hexagon into zonogons are there? The following theorem [Horváth, 1997] provides an answer.

Theorem 7.4.1. *There is an infinite sequence of irreducible dissections of a regular hexagon into finitely many zonogons.*

The proof is based on the dissection into $n(n+5)/2$ zonogons illustrated in Figure 7.4.1c for $n=4$. The dissection consists of $3n$ rhombi and $n(n-1)/2$ regular hexagons and is clearly irreducible.

However, the dissections illustrated by example in Figure 7.4.1c are not edge-to-edge (see Section 1.7). For those we have the following result.

Theorem 7.4.2. *There are exactly six types of edge-to-edge irreducible dissections of a regular hexagon into finitely many zonogons.*

There are two types of such dissections into 3 parts (one of which is the dissection in Figure 7.4.1a), and one dissection each into 4, 6, 8, and 13 parts. See [Horváth, 1997] for a proof. □

The next example presents an area property of convex parahexagons using the fact that they are zonogons that can be dissected into zonogons (in this case rhombi).

7.4. ZONOGONS

Example 7.4.2. *An area property of hexagonal zonogons.* Joining the midpoints of adjacent sides of a hexagonal zonogon creates an inner hexagonal zonogon with 3/4 of the area of the original. See Figure 7.4.2a.

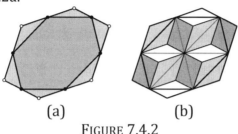

(a)   (b)

FIGURE 7.4.2

To show that this is true, dissect the hexagonal zonogon into 12 rhombi, subdivide each rhombus into two triangles, and shade as shown in Figure 7.4.2b. Similarly shaded triangles are congruent. The outer hexagonal zonogon contains eight white, eight light gray, and eight dark gray triangles, while the inner hexagonal zonogon contains six white, six light gray, and six dark gray triangles, which establishes the result [Bogomolny, 2018]. □

To conclude this section we prove a general result about dissecting zonogons into rhombi [Ball and Coxeter, 1974].

Theorem 7.4.3. *For all $n \geq 3$ every $2n$-sided zonogon, in particular every regular $2n$-gon, can be dissected into $n(n-1)/2$ rhombi whose sides are parallel to and have the same lengths as a pair of sides of the $2n$-sided zonogon.*

FIGURE 7.4.3

Figure 7.4.3 illustrates a dissection of a regular dodecagon into 15 rhombi.

Proof. The proof is by induction on n. For $n = 3$ the dark gray hexagonal zonogon in the lower left corner of Figure 7.4.3 illustrates a dissection with three rhombi. For the induction step, assume the statement is true for any $2n$-sided zonogon, and observe that an arbitrary $2(n+1)$-sided zonogon can be derived from an appropriate $2n$-sided zonogon by adding a "ribbon" of n rhombi as illustrated in Figure 7.4.3 by the rhombi in different shades of gray. In a $2n$-sided zonogon the n pairs of parallel sides have n different orientations, and there is a rhombus in the dissection for each pair of orientations, hence there are $\binom{n}{2} = n(n-1)/2$ rhombi. ∎

7.5. Cyclic polygons

A polygon is *cyclic* if it has a circumcircle, i.e., a circle passing through all the vertices. As noted in Chapter 1 all regular polygons are cyclic, as are all triangles. Many other polygons are also cyclic, and we devote this section to presenting some of their properties.

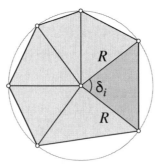

FIGURE 7.5.1

Example 7.5.1. *Area formulas for cyclic polygons.* For an irregular cyclic polygon we have a formula for its area K in terms of the circumradius R and the central angles δ_i. Drawing radii to the vertices partitions the cyclic n-gon into n isosceles triangles as shown in Figure 7.5.1, yielding

(7.1) $$K = \frac{R^2}{2}(\sin\delta_1 + \sin\delta_2 + \cdots + \sin\delta_n).$$

If we place a cyclic n-gon on a coordinate plane so that its circumcenter is at the origin and the vertices are given by $A_i = (x_i, y_i)$, then the expression for the area of a triangle in terms of the absolute value of a determinant yields

(7.2) $$K = \frac{1}{2}\left(\left\|\begin{matrix}x_1 & y_1\\ x_2 & y_2\end{matrix}\right\| + \left\|\begin{matrix}x_2 & y_2\\ x_3 & y_3\end{matrix}\right\| + \cdots + \left\|\begin{matrix}x_n & y_n\\ x_1 & y_1\end{matrix}\right\|\right).$$

The above area formulas for cyclic polygons require some knowledge of angle measures or location of the vertices. There are concise formulas for the area in terms of side lengths for triangles (Heron's formula) and for cyclic quadrilaterals (Brahmagupta's formula), but comparable formulas for cyclic n-gons for $n \geq 5$ are rather more complex. See [Robbins, 1994, 1995] for the cases of cyclic pentagons and hexagons. □

We present several theorems that illustrate the importance and role of cyclic polygons.

Theorem 7.5.1. *An equilateral cyclic polygon is regular.*

Proof. If the cyclic polygon is equilateral, then it can be partitioned into congruent isosceles triangles, so that it is also equiangular and thus regular. ∎

Theorem 7.5.2. *For any simple n-gon, there exists a cyclic n-gon with the same side lengths.*

Proof. Let $a_1, a_2, \cdots, a_n$ be the side lengths of the given n-gon. Let C be a circle with a sufficiently large radius, and starting at an arbitrary point A_1 on C, inscribe chords $A_1A_2, A_2A_3, \cdots, A_nA_{n+1}$ of lengths $a_1, a_2, \cdots, a_n$, respectively. If A_{n+1} does not coincide with A_1 decrease the radius of C continuously until it does, yielding a cyclic n-gon with side lengths $a_1, a_2, \cdots, a_n$. ∎

Although a cyclic n-gon with given side lengths exists, one would need to know its circumcenter or circumradius in order to construct it. See [Garza-Hume, 2018] for an algorithm and computer program to accomplish this.

Theorem 7.5.3. *The maximum area of a simple n-gon with given side lengths occurs when the n-gon is cyclic.*

See [Demir, 1966] for a multivariate calculus proof with Lagrange multipliers, or [Alsina and Nelsen, 2009] for a geometric proof using the isoperimetric theorem.

In the previous two theorems we considered n-gons with given side lengths; in the next theorem we consider cyclic n-gons with a common circumradius.

Theorem 7.5.4. *Among all cyclic n-gons with a given circumradius, the regular n-gon has the largest area.*

Proof. Let K denote the area of a cyclic n-gon with central angles δ_i for $1 \leq i \leq n$ with circumradius R. Applying Jensen's inequality [Needham, 1993] to (7.1) yields

$$(7.3) \qquad K = \frac{R^2}{2}(\sin \delta_1 + \sin \delta_2 + \cdots + \sin \delta_n)$$

$$\leq \frac{nR^2}{2} \sin \frac{\delta_1 + \delta_2 + \cdots + \delta_n}{n} = \frac{nR^2}{2} \sin \frac{2\pi}{n}$$

when all the angles δ_i lie in $[0, \pi]$. If one angle, say δ_k, is greater than π, then we obtain the same inequality by replacing δ_k by π, noting that $\sin \delta_k < \sin \pi$. Since the last term in (7.3) is the area of a regular n-gon inscribed in a circle of radius R, the result follows. ∎

Example 7.5.2. *Equal angle sums in a cyclic 2n-gon.* Proposition 22 in Book III of Euclid's *Elements* states that in a cyclic quadrilateral, the sum of each pair of opposite vertex angles is 180°. An analogous result holds for cyclic $2n$-gons—if the measures of the vertex angles (in order) are $\theta_1, \theta_2, \theta_3, \ldots, \theta_{2n}$, then

$$(7.4) \qquad \theta_1 + \theta_3 + \cdots + \theta_{2n-1} = \theta_2 + \theta_4 + \cdots + \theta_{2n} = (n-1)180°.$$

7.5. CYCLIC POLYGONS

The proof is by induction on n. Euclid proved the base case $n = 2$ in the *Elements*. For the induction step, draw a chord in a cyclic $(2n + 2)$-gon to create a $2n$-gon whose vertex angles satisfy (7.4) and a cyclic quadrilateral with equal opposite angle sums, as indicated by the angle marks ● and ○ in Figure 7.5.2.

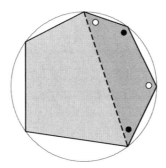

FIGURE 7.5.2

It now follows that 180° is added to each of the angle sums from the $2n$-gon, resulting in equal angle sums for the $(2n + 2)$-gon. □

Example 7.5.3. *Equal alternate side lengths in an equiangular cyclic $2n$-gon.* In the equiangular cyclic $2n$-gon in Figure 7.5.3, $\triangle A_4 A_3 A_2$ is congruent to $\triangle A_1 A_2 A_3$ since $\angle A_4 A_3 A_2 = \angle A_1 A_2 A_3$, $\angle A_2 A_4 A_3 = \angle A_3 A_1 A_2$ (since they intercept the same arc), and have the common side a_2, hence $a_1 = a_3$.

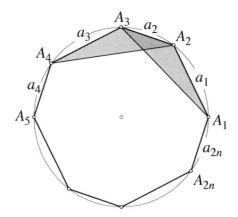

FIGURE 7.5.3

In the same fashion we can show that $a_3 = a_5$ and so on, hence all the sides with odd subscripts have equal length. Similarly $\Delta A_2 A_3 A_4$ is congruent to $\Delta A_3 A_4 A_5$ so that $a_2 = a_4$, and an analogous argument shows that all the sides with even subscripts also have a common length [de Villiers, 2011b]. □

See Challenge 7.7 for a similar result for cyclic $(2n - 1)$-gons.

Example 7.5.4. *The Japanese theorem*. In Section 2.3 we mentioned *sangaku*, tablets with geometry theorems written upon them displayed in Japanese temples and shrines during the Edo period. A classical example from about 1800 is one concerning circles in the interior of a cyclic polygon, known simply as *the Japanese theorem*.

Theorem 7.5.5. *If a polygon is inscribed in a circle and partitioned into triangles by diagonals, then the sum of the inradii of the resulting triangles is a constant independent of the particular triangulation of the polygon.*

In Figure 7.5.4 we have an example showing two different triangulations of a cyclic pentagon. For proofs see [Honsberger, 1985; Johnson, 1960]. □

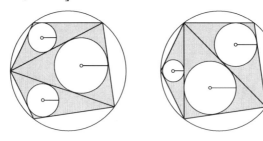

FIGURE 7.5.4

7.6. Tangential polygons

A convex n-gon is *tangential* if it has an incircle tangent to each of its n sides. This is equivalent to specifying that the n vertex angle bisectors are concurrent meeting at a point equidistant from the sides of the n-gon. As noted in Chapter 1 all

7.6. TANGENTIAL POLYGONS

regular polygons are tangential, as are all triangles and all kites (a kite is a convex quadrilateral with side lengths a, a, b, b in that order). We denote the vertices of a tangential n-gon by $A_1, A_2, \ldots, A_n$, side lengths by $a_1 = A_1A_2, a_2 = A_2A_3, \ldots, a_n = A_nA_1$, and inradius by r, as shown in Figure 7.6.1.

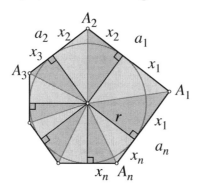

FIGURE 7.6.1

Since two line segments from a point outside a circle tangent to the circle have equal length, we have $a_1 = x_1 + x_2, a_2 = x_2 + x_3, \cdots, a_n = x_n + x_1$, as shown in the figure. It follows that the semiperimeter s of the tangential n-gon in the figure is given by $s = x_1 + x_2 + \cdots + x_n$.

Example 7.6.1. *The area of a tangential polygon.* The inradii and line segments drawn from the incenter to the vertices partition a tangential n-gon into $2n$ right triangles as shown in Figure 7.6.1. The right triangles come in congruent pairs each with area $rx_i/2$; hence

(7.4) $\qquad K = r(x_1 + x_2 + \cdots + x_n) = rs.$ $\square$

Theorems 7.6.1 and 7.6.2 and Examples 7.6.2 and 7.6.3 that follow complement Theorems 7.5.1 and 7.5.4 and Examples 7.5.2 and 7.5.3 for cyclic polygons.

Theorem 7.6.1. *An equiangular tangential polygon is regular.*

Proof. If the tangential n-gon in Figure 7.6.1 is equiangular, then the $2n$ gray right triangles are congruent, so that the n-gon is also equilateral and thus regular. ∎

Theorem 7.6.2. *Among all tangential n-gons with a given inradius, the regular n-gon has the smallest area.*

Proof. Let K be the area of a tangential n-gon as illustrated in Figure 7.6.1. Then by (7.4) and the arithmetic mean-geometric mean inequality we have

$$(7.5) \qquad K = r(x_1 + x_2 + \cdots + x_n) \geq rn\sqrt[n]{x_1 x_2 \cdots x_n}$$

with equality if and only if $x_1 = x_2 = \cdots = x_n$, i.e., when the n-gon is equilateral. But in this case the n-gon is also equiangular (and thus regular) since all the right triangles with legs r and x_i in Figure 7.6.1 are congruent. Note that when (7.5) is an equality, the right side of (7.5) simplifies to rnx_1 where nx_1 is the semiperimeter of the regular n-gon. ∎

Example 7.6.2. *Equal side length sums in a tangential $2n$-gon.* In a tangential $2n$-gon with sides labeled similarly to the one in Figure 7.6.1, we have

$$a_1 + a_3 + \cdots + a_{2n-1} = x_1 + x_2 + x_3 + x_3 + \cdots + x_{2n-1} + x_{2n} = s$$

and similarly

$$a_2 + a_4 + \cdots + a_{2n} = x_2 + x_3 + x_4 + x_5 + \cdots + x_{2n} + x_1 = s.$$

Thus

$$a_1 + a_3 + \cdots + a_{2n-1} = s = a_2 + a_4 + \cdots + a_{2n}. \qquad \square$$

Example 7.6.3. *Equal alternate angles in an equilateral tangential $2n$-gon.* In an equilateral tangential $2n$-gon, $a_1 = a_2 = \cdots = a_{2n}$ implies that $a_i = x_1 + x_2$ for all i from 1 to $2n$, as shown in Figure 7.6.2.

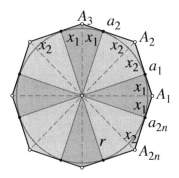

FIGURE 7.6.2

All the right triangles with legs x_1 and r are congruent, as are all the right triangles with legs x_2 and r. Thus the light gray kites are congruent, as are the dark gray kites. Hence the angles at vertices $A_1, A_3, \ldots, A_{2n-1}$ are equal, as are the angles at vertices $A_2, A_4, \ldots, A_{2n}$. □

See Challenge 7.8 for a similar result for tangential $(2n-1)$-gons.

Example 7.6.4. *Side and angle duality.* Examples 7.5.2 and 7.5.3, and Examples 7.6.2 and 7.6.3 suggest a dual relationship between *sides* and *angles* in equiangular cyclic $2n$-gons and equilateral tangential $2n$-gons. To illustrate and summarize this duality [de Villiers, 2011], we consider an equiangular cyclic octagon (a truncated square) and an equilateral tangential octagon (a zonogon) in Figure 7.6.3.

FIGURE 7.6.3

equiangular cyclic 2n-gon	equilateral tangential 2n-gon
equal *angles*	equal *sides*
equal alternate *sides*	equal alternate vertex *angles*
diagonals have equal *length*	diagonals intersect at equal *angles*
axes of symmetry bisect *sides*	axes of symmetry bisect vertex *angles*

The situation for $(2n-1)$-gons is different. An equiangular cyclic $(2n-1)$-gon must also be equilateral; hence it is regular and thus tangential as well. A converse also holds, see Challenges 7.7 and 7.8. □

7.7. Bicentric polygons

A bicentric polygon is one that is both cyclic and tangential, and so-named since it has both an incenter and a circumcenter. All triangles are bicentric, as are regular n-gons and regular star n-gons. For a regular n-gon, the inradius r and circumradius R satisfy $r/R = \cos(\pi/n)$; and for the regular star n-gon $\{n/k\}$ we have $r/R = \cos(k\pi/n)$. Many irregular n-gons are also bicentric. For example, all right kites (kites with a pair of right-angled opposite vertices) are bicentric.

Bicentric polygons have an unexpected and amazing property. Consider a bicentric n-gon with its incircle and circumcircle. Then every point P on the circumcircle is the vertex of another bicentric n-gon, as illustrated in Figure 7.7.1 for an irregular bicentric quadrilateral.

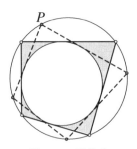

FIGURE 7.7.1

This property is a special case of *Poncelet's closure theorem*, named for the French mathematician and engineer Jean-Victor Poncelet (1788-1867), who gave a proof for polygons inscribed in and circumscribed about conics in 1822. See [Halbeisen and Hungerbühler, 2015] for a modern proof.

Rectilinear farmland

The quote at the beginning of this chapter mentions the polygonal nature of farmland boundaries. Rectilinear layouts are quite common, as seen in Figure 7.7.2. On the left is an aerial view of an agricultural farm in Minnesota, while on the right an aerial view of a "solar farm" in a California desert consisting of solar panels.

FIGURE 7.7.2

7.8. Challenges

7.1 How does the lower bound for the area of the lattice pentagon in Example 7.2.5 change if the word "convex" is deleted from its description?

7.2 Prove that it is impossible for a lattice triangle to be equilateral. (Hint. Assume such a triangle exists and then use Heron's formula and Pick's theorem to compute its area.)

7.3 If two lattice polygons have the same lattice points as vertices, must they be congruent? Must they have the same area?

7.4 Let a be the number of 90° vertices and b the number of 270° vertices in a simple rectilinear polygon. Show that $a = b + 4$.

7.5 Which of the four rectilinear octagons in Figure 7.3.1b tile the plane?

7.6 The area formula (7.2) for cyclic polygons also applies to any polygon in the Cartesian plane with the origin (0,0) as a vertex. Illustrate this fact by using (7.2) to compute the area of the parallel pentagon in Figure 2.5.4.

7.7 Let P be a cyclic $(2n - 1)$-gon. Show that P is equilateral if and only if it is equiangular.

7.8 Let Q be a tangential $(2n - 1)$-gon. Show that Q is equiangular if and only if it is equilateral.

7.9 (i) Suppose an equiangular $2n$-gon has equal alternate side lengths. Must it be cyclic?

(ii) Suppose an equilateral $2n$-gon has equal alternate angles. Must it be tangential?

7.10 An *equable* polygon is one whose area is numerically equal to its perimeter. For example, a right triangle with sides 5, 12, and 13 is equable since its area and perimeter both equal 30, and a 4×4 square is also equable. Are there equable polygons with five or more sides? (Hint. What do the triangle and square in this Challenge have in common?)

CHAPTER 8

Polygonal Numbers

Why are numbers beautiful? It's like asking why is Ludwig van Beethoven's Ninth Symphony beautiful. If you don't see why, someone can't tell you. I know numbers are beautiful. If they aren't beautiful, nothing is.

Paul Erdős

One of the earliest subsets of natural numbers recognized by ancient mathematicians was the set of polygonal numbers. Such numbers represent an ancient link between geometry and number theory.

James J. Tattersall

8.1. Introduction

Polygonal numbers are integer analogs of planar polygons. They belong to a class known as the *figurate numbers*—positive integers that can be represented geometrically by arrangements of points or physically by arrangements of objects like pebbles. Figurate numbers appear in some of the works of early Greek geometers, as well as in contributions by eminent mathematicians such as Jakob Bernoulli, Pierre de Fermat, and Leonhard Euler. See [Deza and Deza, 2011] for a comprehensive overview of figurate numbers.

In this chapter we consider planar figurate numbers known as *polygonal numbers*—numbers represented by arrangements of objects in the plane in regular polygonal patterns. There are two types of polygonal numbers of interest, the original ones that we shall call *ordinary polygonal numbers* (although most sources refer to them simply as *the* polygonal numbers) and less well-known ones called *centered polygonal numbers*. There are

8. POLYGONAL NUMBERS

also planar figurate numbers based on irregular polygons, such as trapezoidal numbers, oblong numbers, etc., but we reserve the adjective "polygonal" for those based on regular polygons. Furthermore, "polygonal number" (e.g., triangular number, square number, etc.) without an adjective will refer to an ordinary polygonal number.

Sequences of ordinary polygonal numbers are generated by constructing regular polygons from a fixed vertex. In Figure 8.1.1 we see polygonal representations of the first four elements in the sequences of ordinary triangular, square, pentagonal, and hexagonal numbers.

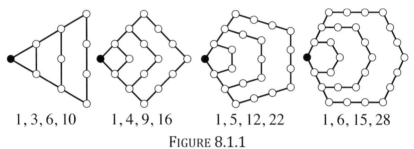

1, 3, 6, 10   1, 4, 9, 16   1, 5, 12, 22   1, 6, 15, 28

FIGURE 8.1.1

The first two images in the above figure show that an ordinary triangular number is the sum of the first few positive integers and that an ordinary square number is the sum of the first few odd integers.

Sequences of centered polygonal numbers are generated by constructing nested regular polygons from a fixed center. In Figure 8.1.2 we see polygonal representations of the first four elements in the sequences of centered triangular, square, pentagonal, and hexagonal numbers.

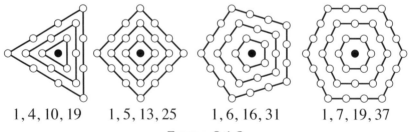

1, 4, 10, 19   1, 5, 13, 25   1, 6, 16, 31   1, 7, 19, 37

FIGURE 8.1.2

8.1. INTRODUCTION

Let $P_{k,n}$ denote the nth ordinary k-gonal number, and let $C_{k,n}$ denote the nth centered k-gonal number. We refer to n as the *rank* and k as the *shape* of both $P_{k,n}$ and $C_{k,n}$.

In Tables 8.1 and 8.2 we display the values of $P_{k,n}$ and $C_{k,n}$ for $3 \leq k \leq 8$ and $1 \leq n \leq 10$.

TABLE 8.1 Ordinary polygonal numbers $P_{k,n}$

$n =$	1	2	3	4	5	6	7	8	9	10
$k = 3$	1	3	6	10	15	21	28	36	45	55
$k = 4$	1	4	9	16	25	36	49	64	81	100
$k = 5$	1	5	12	22	35	51	70	92	117	145
$k = 6$	1	6	15	28	45	66	91	120	153	190
$k = 7$	1	7	18	34	55	81	112	148	189	235
$k = 8$	1	8	21	40	65	96	133	176	225	280

TABLE 8.2 Centered polygonal numbers $C_{k,n}$

$n =$	1	2	3	4	5	6	7	8	9	10
$k = 3$	1	4	10	19	31	46	64	85	109	136
$k = 4$	1	5	13	25	41	61	85	113	145	181
$k = 5$	1	6	16	31	51	76	106	141	181	226
$k = 6$	1	7	19	37	61	91	127	169	217	271
$k = 7$	1	8	22	43	71	106	148	197	253	316
$k = 8$	1	9	25	49	81	121	169	225	289	361

A cursory examination of the two tables reveals some interesting patterns. For example, in Table 8.1 each square number $P_{4,n}$ is the sum of two consecutive triangular numbers, and in Table 8.2 each centered octagonal number $C_{8,n}$ is an odd square. Some numbers appear in both tables in interesting ways, for example, it appears that $C_{k,k}$ equals $P_{k,k+1}$ — does this pattern hold for all $k \geq 3$? We discuss these observations and many others in the next several sections.

8.2. Ordinary polygonal numbers

Table 8.1 yields some valuable information about a general formula for the nth ordinary k-gonal number $P_{k,n}$: it should be linear in k since the numbers in each column increase linearly and quadratic in n since the differences of adjacent numbers in each row increase linearly. To find a formula we "triangulate" $P_{k,n}$ much like the way the diagonals of a regular polygon partition it into triangles. We illustrate the procedure with $P_{6,5} = 45$ in Figure 8.2.1.

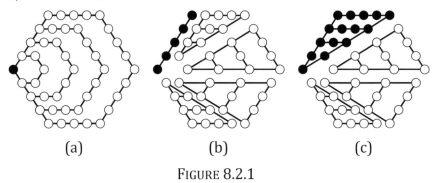

FIGURE 8.2.1

In Figures 8.2.1b and 8.2.1c we see that $P_{6,5} = 5 + 4T_4 = T_5 + 3T_4$, where for convenience we let T_n denote the nth ordinary triangular number $P_{3,n}$. In general we have

(8.1) $\qquad P_{k,n} = n + (k-2)T_{n-1} = T_n + (k-3)T_{n-1}.$

Since $T_n = n(n+1)/2$ (see Challenge 8.1) we see that $P_{k,n}$ is linear in k and quadratic in n as suspected, specifically for $n \geq 1$ and $k \geq 3$,

(8.2) $\qquad P_{k,n} = \frac{1}{2}[(k-2)n^2 - (k-4)n].$

The triangular numbers play a role in the family of polygonal numbers analogous to the role played by triangles in the family of polygons, as the following examples illustrate.

Example 8.2.1. *Sums of arithmetic progressions are polygonal.* One of the problems in Diophantus of Alexandria's *Arithmetica* states (in modern notation) that if $a_k = 1 + (k-1)d$ is an arithmetic progression with first term 1 and common difference d a positive integer, then the sum $S_{d,n}$ of the first n terms is given by

$$S_{d,n} = \sum_{k=1}^{n} a_k = n + dT_{n-1} = P_{d+2,n};$$

that is, $S_{d,n}$ is an ordinary polygonal number with rank n and shape $d + 2$. □

Example 8.2.2. *Pentagonal, trapezoidal, and triangular numbers.* The nth ordinary pentagonal number $P_{5,n}$ can be computed by $n + 3T_{n-1}$ or $T_n + 2T_{n-1}$, but there are two other less obvious but useful ways, as shown in Figure 8.2.2.

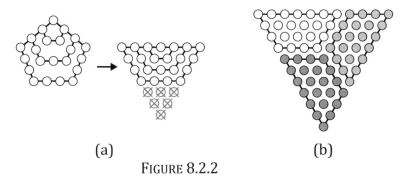

(a) (b)

FIGURE 8.2.2

Distorting the pentagonal shape of $P_{5,n}$ into the trapezoidal shape in Figure 8.2.2a shows that $P_{5,n} = T_{2n-1} - T_{n-1}$, and Figure 8.2.2b shows how three copies of the trapezoidal shape yields a triangular number, i.e., $3P_{5,n} = T_{3n-1}$. Hence $P_{5,n} = T_{3n-1}/3$. □

A *trapezoidal number* is an integer that can be written as the sum of two or more consecutive integers greater than 1. So Figure 8.2.2a shows that every ordinary pentagonal number is a trapezoidal number.

Example 8.2.3. *Hexagonal numbers are triangular.* Examination of Table 8.1 suggests that every ordinary hexagonal number equals a triangular number with odd rank and that their common value is the product of their ranks, i.e., $P_{6,n} = T_{2n-1} = n(2n-1)$. See Challenge 8.1 and see Figure 8.2.2 for an illustration on the cover of the January 2014 issue of the *College Mathematics Journal* for $n = 5$, i.e., $P_{6,5} = T_9 = 5 \cdot 9$. □

 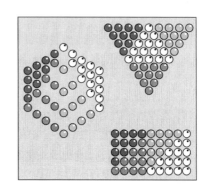

FIGURE 8.2.3

Example 8.2.4. *Even perfect numbers are hexagonal.* A *perfect number* is a positive integer n that is equal to the sum of its *proper divisors* (those divisors less than n). For example, 6 is perfect since $6 = 1 + 2 + 3$, and 28 is perfect since $28 = 1 + 2 + 4 + 7 + 14$. Perfect numbers appear in Definition 22 in Book VII of Euclid's *Elements* [Heath, 1956] as "*A perfect number is that which is equal to its own parts.*" Euclid also gives a formula for even perfect numbers in terms of a prime p (Proposition 36 in Book IX of the *Elements*), which in modern notation reads: *If p and $q = 2^p - 1$ are prime, then $N_p = 2^{p-1}q$ is perfect.* Only 51 perfect numbers are known, and the first six are 6, 28, 496, 8128, 33550336, and 8589869056.

Since $N_p = 2^{p-1}q = 2^p(2^p-1)/2$ we see that $N_p = T_{2^p-1}$, and from the preceding example we have $N_p = T_{2^p-1} = P_{6,2^{p-1}}$, so that even perfect numbers are both triangular and hexagonal. For example, when $p = 5$, $2^5 - 1 = 31$, and $N_5 = T_{31} = P_{6,16} = 16 \cdot 31 = 496$. Two ancient open questions are: do infinitely many even perfect numbers exist, and are there any odd perfect numbers? □

Example 8.2.5. *Square triangular numbers.* A cursory examination of Table 8.1 shows that 1 and 36 are square triangular numbers. Are there others? To show that there are infinitely many, we first note that an odd square is 1 more than 8 times a triangular number, i.e., $(2k+1)^2 = 4k^2 + 4k + 1 = 8T_k + 1$. Replacing n in $T_n = n(n+1)/2$ with $8T_k$ and simplifying yields

$$T_{8T_k} = \frac{8T_k(8T_k+1)}{2} = 4T_k(2k+1)^2.$$

So if T_k is a square, so is T_{8T_k}. Since $T_1 = 1^2$, we have $T_8 = 36 = 6^2$, $T_{288} = 41616 = 204^2$, and so on. But this only generates even square triangular numbers, and odd square triangular numbers also exist, such as $T_{49} = 35^2$, $T_{1681} = 1189^2$, etc. For details and other *multi-polygonal numbers* (numbers polygonal in two or more ways), see [Deza and Deza, 2012]. □

Example 8.2.6. *Polygonal numbers in Pascal's triangle.* The reader is no doubt familiar with Pascal's triangle, a triangular arrangement of positive integers with a variety of nice properties. The elements of the triangle are the *binomial coefficients* $\binom{n}{j} = \frac{n(n-1)(n-2)\cdots(n-j+1)}{j(j-1)(j-2)\cdots 2 \cdot 1}$ for integers $n \geq 0$ and $0 \leq j \leq n$. See Figure 8.2.4 for the first nine rows (row $n = 0$ to row $n = 8$ inclusive).

```
              1
            1   1
          1   2   1
        1   3   3   1
      1   4   6   4   1
    1   5  10  10   5   1
  1   6  15  20  15   6   1
1   7  21  35  35  21   7   1
1 8 28 56 70 56 28 8  1
```

FIGURE 8.2.4

Of interest are the appearances of some polygonal numbers in Pascal's triangle. For example, all the triangular numbers appear in the $k = 2$ diagonal of the triangle (the third element of each row starting with row $n = 2$) since $\binom{n}{2} = \frac{n(n-1)}{2} = T_{n-1}$ for $n \geq 2$.

The following theorem and corollary explain why some pentagonal numbers from Table 8.1 appear in the $k = 4$ diagonal (the fifth element of each row starting with row $n = 4$), e.g., $P_{5,1} = 1 = \binom{4}{4}$, $P_{5,2} = 5 = \binom{5}{4}$, $P_{5,5} = 35 = \binom{7}{4}$, $P_{5,7} = 70 = \binom{8}{4}$; as well as some not shown in Table 8.1 or Figure 8.2.4, e.g., $P_{5,12} = 210 = \binom{10}{4}$, $P_{5,15} = 330 = \binom{11}{4}$, etc.

Theorem 8.2.1. *A pentagonal number with pentagonal rank is a binomial coefficient in the $k = 4$ diagonal of Pascal's triangle, that is, for $n \geq 1$,*

$$P_{5,P_{5,n}} = \binom{3n+1}{4}.$$

Proof. Since $P_{5,n} = n(3n-1)/2$ we have

$$P_{5,P_{5,n}} = \frac{1}{2}P_{5,n}(3P_{5,n} - 1) = \frac{1}{8}n(3n-1)(9n^2 - 3n - 2)$$

$$= \frac{1}{24}(3n)(3n-1)(3n+1)(3n-2) = \binom{3n+1}{4}. \blacksquare$$

Since $P_{5,n} = T_{3n-1}/3$ (see Example 8.2.2) pentagonal numbers with rank $T_{3n}/3 = n(3n+1)/2$ may also be binomial coefficients in the $k = 4$ diagonal. That is indeed the case.

Corollary 8.2.2. *A pentagonal number with rank $T_{3n}/3$ is also a binomial coefficient in the $k = 4$ diagonal of Pascal's triangle, that is, for $n \geq 1$,*

$$P_{5,T_{3n}/3} = \binom{3n+2}{4}.$$

The proof is analogous to that of the theorem and is omitted.

With all the triangular numbers in the $k = 2$ diagonal and some pentagonal numbers in the $k = 4$ diagonal, do some or all of the squares appear in the $k=3$ diagonal? Yes, but only three: $1^2 = \binom{3}{3}$, $2^2 = \binom{4}{3}$, and $140^2 = \binom{50}{3}$ [Dickson, 1966]. Furthermore, the hexagonal numbers appear the $k=2$ diagonal (rather than the $k=5$ diagonal) since every hexagonal number is triangular (see Example 8.2.3). □

8.3. Centered polygonal numbers

To derive a formula for $C_{k,n}$ we again "triangulate" the polygonal representation, this time analogous to drawing radii from the center of a regular n-gon to the n vertices. We illustrate the procedure with $C_{6,5}$ in Figure 8.3.1ab.

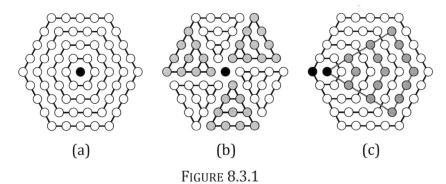

FIGURE 8.3.1

Figure 8.3.1b shows that $C_{6,5} = 1 + 6T_4$, and in general we have

(8.3) $$C_{k,n} = 1 + kT_{n-1}.$$

The centered hexagonal numbers illustrated in Figure 8.3.1 are often simply called *hex numbers* and are denoted by $H_n = C_{6,n}$.

Example 8.3.1. *The relationship between $P_{k,n}$ and $C_{k,n}$.* Tables 8.1 and 8.2 suggest that the difference between $C_{k,n}$ and $P_{k,n}$ is a square that depends only on the rank n and not on the shape k, in fact for $n \geq 1$ and $k \geq 3$ we have

$$C_{k,n} = P_{k,n} + (n-1)^2.$$

See Figure 8.3.1c for an illustration of the case $C_{6,5} = P_{6,5} + 4^2$. The algebraic proof for the general case is simple. From (8.1) and (8.3) we have

$$C_{k,n} - P_{k,n} = 1 + kT_{n-1} - [n + (k-2)T_{n-1}]$$
$$= 2T_{n-1} - (n-1) = (n-1)^2. \square$$

Example 8.3.2. *Numbers both ordinary and centered n-gonal.* In Tables 8.1 and 8.2 it is easy to find equal ordinary polygonal and centered polygonal numbers with the same shape. They appear in the two tables in the downward-sloping diagonals with elements 10, 25, 51, etc., where $P_{n,n+1} = C_{n,n}$. There are many others not in the tables, e.g., $P_{3,16} = 136 = C_{3,10}$, $P_{4,29} = 841 = C_{4,21}$, and infinitely many more. For details see [Schlicker, 2011]. $\square$

The second quote at the beginning of this chapter mentions polygonal numbers as an ancient link between geometry and number theory. One such link concerns centered polygonal numbers and *Pythagorean triples*—triples (a,b,c) of positive integers with $a^2 + b^2 = c^2$ so that a, b, and c can be side lengths in a right triangle.

Example 8.3.3. *Square centered square numbers and almost isosceles Pythagorean triples.* When a and b differ by 1 in a Pythagorean triple (a,b,c), it is called *almost isosceles*. Now suppose $C_{4,n} = m^2$, as in $C_{4,4} = 5^2$ and $C_{4,21} = 29^2$. Since $C_{4,n} = n^2 + (n-1)^2$ (see Challenge 8.2) we have the two almost isosceles Pythagorean triples (3,4,5) and (20,21,29). Are there more almost isosceles Pythagorean triples in the sequence of centered square numbers, that is, more square centered square numbers? The answer is yes, infinitely many more, and they are obtained by solving the Diophantine equation $n^2 + (n-1)^2 = m^2$ for n and m. The first few solutions are $(n,m) = (4,5)$, (21,29), (120,169), and (697,985). $\square$

8.4. Other figurate numbers derived from polygons

In this section we present several examples of figurate numbers based on polygons other than regular n-gons.

Example 8.4.1. *Trapezoidal numbers.* In Example 8.2.1 we mentioned *trapezoidal numbers*—numbers that are the sum of two or more consecutive integers greater than one, and we showed that ordinary pentagonal numbers are trapezoidal. Many other integers are trapezoidal, e.g., every odd number 5 or greater is trapezoidal, since $2n + 1 = n + (n + 1)$. Trapezoidal numbers belong to a class known as the *polite numbers*, numbers that are triangular or trapezoidal (or both), e.g., $15 = 7 + 8 = 4 + 5 + 6 = 1 + 2 + 3 + 4 + 5$. It is easy to show [Gamer et al. 1985] that a positive integer is polite if and only if it is not a power of 2. However, the non-trapezoidal polite numbers (i.e., triangular numbers that are not trapezoidal) are relatively rare [Jones and Lord, 1999]; the only ones less than 2,000,000,000 are

3, 6, 10, 28, 136, 496, 8128, 32896, and 33550336. □

Example 8.4.2. *Star numbers.* A *star number* is the numeric version of a *hexagram*, the star hexagon from Section 3.8. The nth star number S_n has a geometric representation consisting of a central dot surrounded by 12 copies of T_{n-1} dots, so $S_n = 1 + 12T_{n-1} = 1 + 6n(n-1)$, as shown in Figure 8.4.1a for $n = 5$.

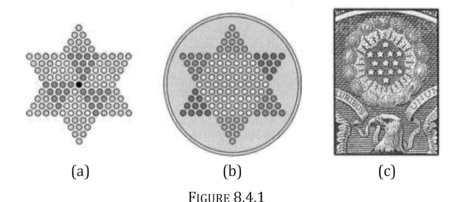

(a) (b) (c)

FIGURE 8.4.1

The star number $S_5 = 121$ is also the layout of the gameboard for the game of Chinese checkers, as illustrated in Figure 8.4.1b. The second star number $S_2 = 13$ appears on the green side of the American one-dollar bill, as shown in Figure 8.4.1c.

The star number S_n is equal to $C_{12,n}$, the nth centered dodecagonal number, although the geometric representations of the two numbers are different. The first 12 star numbers are

$$1, 13, 37, 73, 121, 181, 253, 337, 433, 541, 661, \text{ and } 793.$$

The first and fifth star numbers are squares: $S_1 = 1^2$ and $S_5 = 11^2$. There are infinitely many more square star numbers, found by solving the Diophantine equation $1 + 6n(n-1) = m^2$. The first 5 square star numbers are

$$1, 121 = 11^2, 11881 = 109^2, 1164241 = 1079^2, \text{ and}$$
$$114083761 = 10681^2.$$

Square star numbers have an interesting property: if S_n is a square, then $3S_n + 2$ is both a sum of two consecutive squares and a sum of three consecutive squares. For example, $S_5 = 11^2 = 121$ and $365 = 13^2 + 14^2 = 10^2 + 11^2 + 12^2$. See Challenge 8.6. □

Polygonal numbers in the history of mathematics

In a quote at the beginning of this chapter we noted that the polygonal numbers link geometry and number theory. A case in point is the beautiful theorem, first stated by Pierre de Fermat (1601-1665) in a letter to Blaise Pascal (1623-1662) on September 25, 1654, that every positive integer is the sum of three or fewer triangular numbers, four or fewer squares, five or fewer pentagonal numbers, and in general n or fewer n-gonal numbers. His proof (if it existed) has never been found. Carl Friedrich Gauss proved the triangular case and on July 10, 1796 wrote in his diary "EYPHKA! Num = Δ + Δ + Δ." The case for squares was proven by 1770 by Joseph-Louis Lagrange (1736-1813) and is known as Lagrange's four-squares theorem. In

1813 Augustin-Louis Cauchy (1789-1857) proved the general case of Fermat's claim. Images of Gauss appear in Chapters 1 and 7; here are portraits of Pascal, Fermat, Cauchy, and Lagrange (from left to right).

Figure 8.4.2

8.5. Challenges

8.1 Prove the following results about ordinary polygonal numbers: For $n \geq 1$,

(i) $T_n = n(n+1)/2$; (ii) $T_{n-1} + T_n = n^2$;

(iii) $P_{5,n} = n^2 + T_{n-1}$; (iv) $P_{6,n} = T_{2n-1}$;

(v) $P_{8,n} = (2n-1)^2 - (n-1)^2$;

(vi) $P_{k,n} = n^2 + (k-4)T_{n-1}$.

8.2 Prove the following results about centered polygonal numbers: For $n \geq 1$,

(i) $C_{3,n} = T_{n-2} + T_{n-1} + T_n$ (for $n \geq 3$);

(ii) $C_{4,n} = n^2 + (n-1)^2$; (iii) $C_{6,n} = n^3 - (n-1)^3$;

(iv) $C_{8,n} = (2n-1)^2$; (v) $C_{9,n} = T_{3n-2}$;

(vi) $C_{n,n+2} = C_{n+2,n+1}$; (vii) $C_{4,n} + 4T_{n-1} = (2n-1)^2$;

(viii) $C_{6,n} + 3T_{n-1} = T_{3n-2}$.

8.3 In Example 8.2.1 we showed that $P_{5,n} = P_{3,2n-1} - P_{3,n-1}$, and in Challenge 8.1 you showed that $P_{8,n} = P_{4,2n-1} - P_{4,n-1}$. Now show that for any $k \geq 3$ and $n \geq 2$ we have
$$P_{3k-4,n} = P_{k,2n-1} - P_{k,n-1}.$$

8.4 Show that for all $k \geq 3$ and $n \geq 1$,
$$(2k-1)P_{2k+1,n} + T_{k-2} = T_{(2k-1)n-(k-1)},$$
so that $5P_{7,n} + 1 = T_{5n-2}$, $7P_{9,n} + 3 = T_{7n-3}$, and so on.

8.5 (i) Show that the product of the nth odd square and the nth centered square number is always triangular, e.g., for $n = 3$, $5^2 \cdot 13 = 325 = T_{25}$.

(ii) Show that the product of a hex number (i.e., a centered hexagonal number) H_n and a star number S_n of the same rank is always a triangular number, e.g., for $n = 3$, $H_3 S_3 = 19 \cdot 37 = 703 = T_{37}$.

(iii) Generalize. (Hint. (i) and (ii) yield $C_{4,n} C_{8,n} = T_{C_{8,n}}$ and $C_{6,n} C_{12,n} = T_{C_{12,n}}$.)

8.6 Let S_n be a square star number. Show that $3S_n + 2$ is both a sum of two consecutive squares and a sum of three consecutive squares. (Hint. Multiply both sides of $1 + 6n(n-1) = m^2$ by 3, add 2, and express the left side as a sum of two squares and the right side as a sum of three squares.)

8.7 In Table 8.1 it appears that the only prime ordinary polygonal numbers are numbers of the form $P_{k,2}$ for k a prime. Prove that this is indeed the case.

8.8 In Example 8.2.4 we showed that every even perfect number is both (ordinary) triangular and hexagonal. Show that every even perfect number greater than 6 is also a centered nonagonal number.

8.9 The third star number 37 is also the fourth hex number, i.e., $S_3 = 37 = H_4$, as shown in Figure 8.5.1. Are there other "star hexes"?

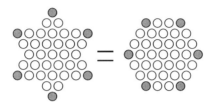

Figure 8.5.1

Solutions to the Challenges

Chapter 1

1.1 From the proof of Theorem 1.4.1 we have
$$\frac{p_n}{P_n} = \frac{\sin(\pi/n)}{\tan(\pi/n)} = \cos\frac{\pi}{n} \text{ and } \frac{k_n}{K_n} = \frac{\sin(\pi/n)\cos(\pi/n)}{\tan(\pi/n)} = \cos^2\frac{\pi}{n}.$$

1.2 The perimeters p_n and P_n satisfy $p_n = na_n$ and $P_n = nA_n$ so that from (1.2) we have
$$2nA_{2n} = \frac{2na_n \cdot nA_n}{na_n + nA_n} \text{ so that } A_{2n} = \frac{a_n A_n}{a_n + A_n} \text{ and}$$
$$2na_{2n} = \sqrt{na_n \cdot 2nA_{2n}} \text{ so that } a_{2n} = \sqrt{a_n A_{2n}/2}.$$

1.3 From the data in Section 1.4 we have $\sin(\pi/2n) = a_{2n}/2\rho$ and $\cos(\pi/2n) = a_n/2a_{2n}$. Now use $\sin^2 x + \cos^2 x = 1$ and solve a quadratic for a_{2n}^2 and take square roots. Similarly we have $\tan(\pi/2n) = A_{2n}/2\rho$ and $\tan(\pi/n) = A_n/2\rho$. Then use the identity $2\cot 2x = \cot x - \tan x$ to solve a quadratic for $1/A_{2n}$ and take reciprocals.

1.4 Using the hint produces the tilings indicated in Figure S1.1. When the triangles are adjacent there are also vertices where six triangles meet, and when the triangles are not adjacent there are also vertices where two dodecagons and one triangle meet. Such tilings are called "2-uniform."

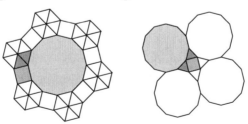

Figure S1.1

1.5 See Figure S1.2. This pentagonal tiling is known as the prismatic pentagonal tiling. As in the Cairo tiling in Figure 1.8.3, the pentagons have two 90° angles and three 120° angles, but here the two 90° angles are adjacent.

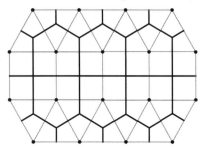

Figure S1.2

1.6 For a convex n-gon the sum of the exterior angles equals $n180°$ minus the sum $(n-2)180°$ of the interior angles, i.e., 360°. For a visual proof, see Figure S1.3, where each angle drawn in the center of the polygon is congruent to one of the exterior angles..

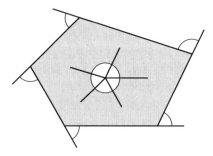

Figure S1.3

1.7 If a convex polygon has four or more acute angled vertices, then there are four or more obtuse exterior angles, which is impossible since the exterior angles sum to 360°. If a convex polygon has seven or more sides, then it has more obtuse angled vertices than acute angled vertices, so at least one side must connect two obtuse vertices.

1.8 Using the hint, let APQ and AQR be two adjacent triangles a shown in Figure S1.4a, with side lengths a, b, c, x, y as shown in Figure S1.4b. Since $AB \cdots N$ is a regular n-gon, the angles

PAQ and *QAR* (marked •) are equal. In fact, all the angles at *A* are equal.

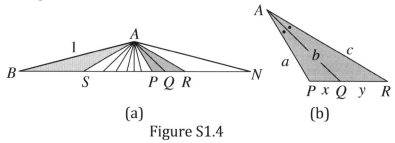

Figure S1.4

Hence *AQ* bisects ∠*PAR* and by the angle bisector theorem we have $x/y = a/c$, or equivalently $a/x = c/y$. Multiplying by b yields $ab/x = bc/y$, so that for each of the small triangles, the product of the lengths of the two sides that meet at *A* divided by the length of the side on *BN* is constant, call it k.

Now consider $\triangle ABS$ in Figure S1.4a. Since $\angle ABS = \angle BAS$ this triangle is isosceles with $BS = AS$. For the constant k we now have
$$k = \frac{AB \cdot AS}{BS} = AB = 1,$$
and thus $ab = x$, $bc = y$, etc. [Koether and Kay, 1978].

1.9 The ratio of the two side lengths is the same as the ratio of the two inradii, so that the side length of the inner *n*-gon is
$$a\frac{x}{r} = a\frac{R\cos((n-1)\pi/2n)}{R\cos(\pi/2n)} = a\frac{\sin(\pi/2n)}{\cos(\pi/2n)} = a\tan\frac{\pi}{2n}.$$

Chapter 2

2.1 Since $AB = AE = \sqrt{3-\varphi}$ and $AC = AD = \varphi\sqrt{3-\varphi}$, the product equals
$$\varphi^2(3-\varphi)^2 = (3\varphi - \varphi^2)^2 = (2\varphi - 1)^2 = (\sqrt{5})^2 = 5.$$

2.2 Let x and y be the lengths of the legs of the right triangle clipped off, as shown in Figure S2.1 [Ren, 2014]. Then

$$a^2 + b^2 + c^2 = a^2 + b^2 + (x^2 + y^2)$$
$$= (a^2 + x^2) + (b^2 + y^2) = d^2 + e^2.$$

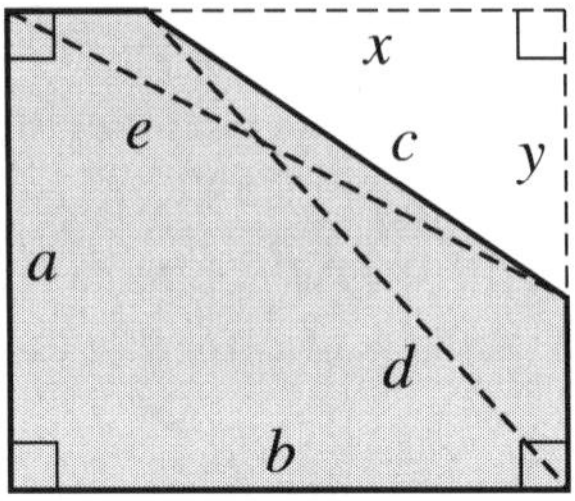

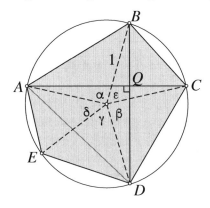

Figure S2.1

2.3 Draw radii to the five vertices, and label the angles at the center as shown in Figure S2.2. The area of P in terms of the five isosceles triangles is

$$\frac{1}{2}\sin\alpha + \frac{1}{2}\sin\beta + \frac{1}{2}\sin\gamma + \frac{1}{2}\sin\delta + \frac{1}{2}\sin\varepsilon.$$

Figure S2.2

Since $\alpha = 2\angle ADQ$ and $\beta = 2\angle DAQ$, α and β are supplementary, so that $\sin\alpha = \sin\beta$. Thus $\sin\alpha + \sin\beta$ attains a maximum at $\alpha = \beta = \pi/2$. Then $\gamma + \delta + \varepsilon = \pi$, so these angles form a triangle. It is well known that that $\sin\gamma + \sin\delta + \sin\varepsilon \leq 3\sqrt{3}/2$ with equality if and only if the triangle is equilateral. Hence $\sin\gamma + \sin\delta + \sin\varepsilon$ attains a maximum at $\gamma = \delta = \varepsilon = \pi/3$. Thus the maximum area of the pentagon is $(2 \cdot 1 + 3 \cdot \sqrt{3}/2)/2 = 1 + (3\sqrt{3}/4)$.

2.4 Since any polygon can be tiled by triangles, it suffices to show that any triangle can be tiled with convex pentagons. Given any two triangles in the plane, one can be converted to the other by an affine transformation (a transformation that preserves linearity and parallelism). Such a transformation sends convex pentagons into convex pentagons, so it suffices to exhibit one triangle that can be tiled by convex pentagons. Here is one [Krusemeyer et al., 2012]:

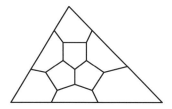

Figure S2.3

2.5 Let M be the intersection of AD and CE as shown in Figure S2.4, and let $[X]$ denote the area of polygon X. Since $ABCDE$ is a parallel pentagon, each of the shaded triangles has area 1. If we let $[AEM\} = x$, then $[ABCDE] = 3 + x$. Since $[CDM] = [AEM] = x$ we have

$$[AEM\}/[DEM] = AM/DM = [ACM]/[CDM],$$

or in terms of x, $x/(1-x) = 1/x$. Hence $x^2 + x - 1 = 0$, so $x = (\sqrt{5} - 1)/2 = \varphi - 1$ and $[ABCDE] = 2 + \varphi$.

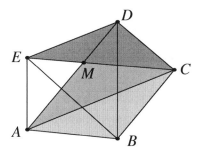

Figure S2.4

2.6 Re-label Figure 2.11.2 as shown in Figure S2.5. Then $BM = BC = 1/\varphi^2$, $CN = CD = 1/\varphi^4$, etc., and $AB + BC + CD + \cdots = AZ$, which proves the claim [Edwards, 2014].

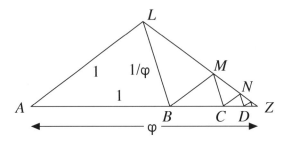

Figure S2.5

2.7 (i) Joining each vertex to the center partitions the pentagon into five congruent isosceles triangles, hence the pentagon is equilateral and equiangular, and thus regular.

(ii) Label the pentagon as $ABCDE$ (in that order). Then $A = B$ implies $\operatorname{arc} BCDE = \operatorname{arc} CDEA$ and hence $\operatorname{arc} BC = \operatorname{arc} EA$, so that $BC = EA$. Similarly $CD = AB$, $DE = BC$, and $EA = CD$. Hence $AB = CD = EA = BC = DE$ and the pentagon is also equilateral and thus regular.

2.8 Make two copies $B_1C_1D_1E_1F_1$ and $B_2C_2D_2E_2F_2$ of $BCDEF$, and orient them so that B_1 and C_2 coincide at P, F_1 and D_2 coincided at Q, and E_1 and E_2 coincide at R, as shown in Figure S2.6.

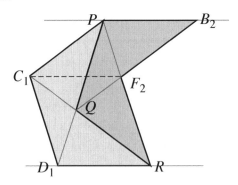

Figure S2.6

Using the fact that each copy of $BCDEF$ consists of two golden triangles and a golden gnomen, it is easy to show that both B_2PC_1F and $F_2C_1D_1R$ are parallelograms. Thus the convex hexagon $B_2PC_1D_1RF_2$ can tile an infinite strip. Hence it can tile the entire plane, and it follows that so can $BCDEF$.

2.9 (i) Assume that the base and height of the pentagon are each 2 so that its area is 3. When the two small isosceles triangles are cut off and moved to the fill the "V" of the pentagon, the result is a 3/2 by 2 rectangle. Then the zigzag cut multiplies the height by 7/6 and the width by 6/7, producing a 7/4 by 12/7 rectangle, rather than a $\sqrt{3}$ by $\sqrt{3}$ square.

(ii) See Figure S2.7. Other solutions may be possible.

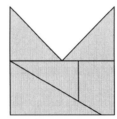
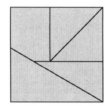

Figure S2.7

2.10 See Figure S2.8.

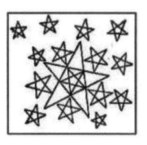

Figure S2.8

Chapter 3

3.1 Since the angles each equal 120° we can extend the sides to form an equilateral triangle as shown in Figure S3.1a. Then $a + b + c = c + d + e$, or $a - d = e - b$. Similarly $e + f + a = a + b + c$, or $e - b = c - f$. Hence $a - d = e - b = c - f$.

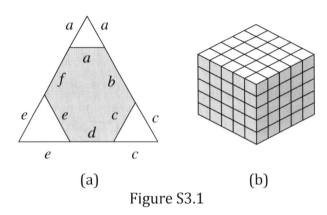

(a)   (b)

Figure S3.1

3.2 See Figure S3.1b, which resembles three square faces of a cube.

3.3 Let $AC = CE = EA = s$, as shown in Figure S3.2. Applying Lemma 3.2.2 to D and $\triangle ACE$ yields $AD = CD + DE$, to B and $\triangle ACE$ yields $BE = AB + BC$, and to F and $\triangle ACE$ yields $CF = EF + FA$, and thus the sum of the long diagonals equals the perimeter of the hexagon.

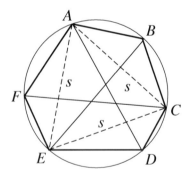

Figure S3.2

3.4 The converse is false. See Figure S3.3 (a modified version of Figure 3.3.1) where $A + C' + E = B + D' + F = 2\pi$ but $ABC'D'EF$ is not cyclic [De Villiers, 2016].

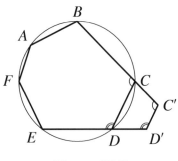

Figure S3.3

3.5 (i) The circumradius R of the hexagon in Figure 3.3.2a is the same as for the hexagon in Figure S3.4a, so it is the circumradius of the equilateral triangle with side s. Since $\cos(2\pi/3) = -1/2$, the law of cosines yields $s^2 = x^2 + y^2 + xy$ and hence $R = (\sqrt{3}/3)\sqrt{x^2 + xy + y^2}$.

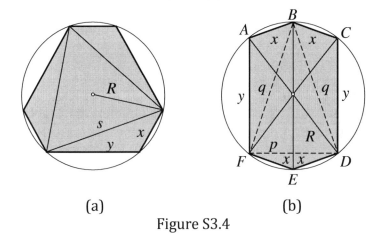

(a) (b)

Figure S3.4

(ii) Rearrange the sides of the hexagon in Figure 3.9.1 to yield $ABCDEF$ in Figure S3.4b; let $p = FD$ and let $q = BD = BF$. Applying Ptolemy's theorem to $ABDF$ yields $2Rq = xp + yq$, so that $xp/q = 2R - y$. Applying Ptolemy's theorem to $BDEF$ yields $2Rp = 2xq$, so that $xq/p = R$. Hence

$$x^2 = (xq/p)(xp/q) = 2R^2 - yR,$$

and so R is the positive root of $2R^2 - yR - x^2 = 0$, i.e., $R = (y + \sqrt{y^2 + 8x^2})/4$.

3.6 (i) See Figure S3.5a. (ii) See Figure S3.5b [De Villiers, 2011].

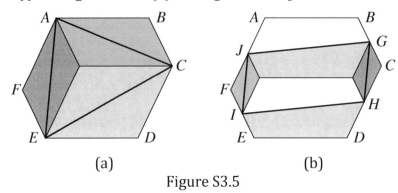

(a) (b)

Figure S3.5

3.7 Apply the triangle inequality to triangles in the parahexagon in Figure 3.5.3b whose sides are the dashed diagonals and adjacent sides of the parahexagon yields $2m_a \leq b+c$, $2m_b \leq a+c$, and $2m_c \leq a+b$, so that $2(m_a + m_b + m_c) \leq 2(a+b+c)$. Applying this inequality to the triangle whose side length are $2m_a, 2m_b$, and $2m_c$ and medians of lengths $3a/2$, $3b/2$, and $3c/2$ yields $3(a+b+c)/2 \leq 2(m_a + m_b + m_c)$, which completes the proof.

3.8 Rather than dissect an L-polyomino into copies of itself, we construct a larger similar one. Consider an L-polyomino with n unit squares. Two copies form a $2\times n$ rectangle, and $2n$ copies of the rectangle form a $2n\times 2n$ square. Finally, n such squares can be joined to yield a large L-polyomino similar to the original one.

3.9 See Figure S3.6.

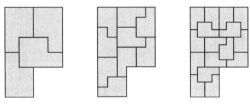

Figure S3.6

3.10 In Figure S3.7a we have overlaid a grid of small triangles similar to the original, showing that a triangle inside the hexagon has 1/16 the area of the original triangle. In

Figure S3.7b we see an enlarged version of the hexagon, overlaid with another grid of similar triangles. Counting small triangles shows that the irregular hexagram has 28/25 the area of the gray triangle in Figure S3.7a. The area of each of the six white triangles in the hexagon is twice that of a small triangle in the grid, and thus the area of the hexagon is $(28+12)/25 = 8/5$ of the gray triangle in Figure S3.7a. Hence the area of the hexagon is $(8/5) \cdot (1/16) = 1/10$ of the original triangle.

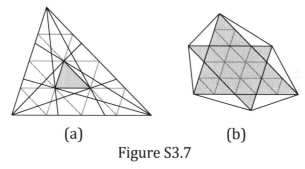

Figure S3.7

3.11 First note that the area of each of the four white triangles in Figure 3.11.4 is $ab/2$, the area of the central right triangle. The triangle between the squares with sides a and b is congruent to the central triangle, and rotating the other two 90° as in Figure S3.8 shows they each have the same base and altitude as the central triangle.

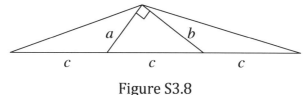

Figure S3.8

Hence the area of the hexagon is $a^2 + b^2 + c^2 + 4(ab/2) = (a+b)^2 + c^2$.

3.12 Using the hint we consider right triangles $a^2 + b^2 = c^2$ with $(a,b,c) = (7,24,25)$ and $(15,20,25)$ to construct the hexagon in Figure S3.9.

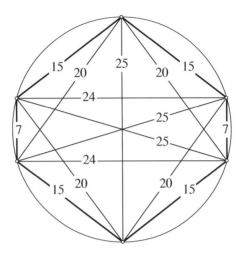

Figure S3.9

Other examples include

$$(a, b, c) = (16,63,65) \text{ and } (25,60,65)$$

and

$$(a, b, c) = (13,84,85) \text{ and } (75,40,85).$$

Infinitely many more such hexagons exist; see [Wickens, 1944].

3.13 Yes, here are two examples [Vaderlind, 2002].

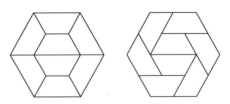

Figure S3.10

3.14 Let h_i denote the altitude of triangle with area T_i in the hexagon from Figure 3.11.5, which we have included in Figure S3.11 below for convenience.

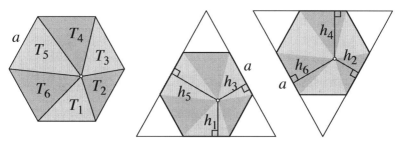

Figure S3.11

Extend three sides of the hexagon in two different ways to circumscribe a pair of equilateral triangles about the hexagon. From Theorem 1.3.2 we have

$$h_1 + h_3 + h_5 = 3r = h_2 + h_4 + h_6,$$

where r is the common inradius of the hexagon and the equilateral triangles. Multiplying by $a/2$ and recalling that $K = 3ar$ yields

$$T_1 + T_3 + T_5 = K/2 = T_2 + T_4 + T_6.$$

This result readily extends to regular $2n$-gons by circumscribing regular n-gons in two ways.

3.15 Let $\hat{\hat{\pi}}$ denote the approximating value of π. Then $6/2\hat{\hat{\pi}} = 57/60 + 36/3600 = 24/25$, and thus $\hat{\hat{\pi}} = 25/8$.

3.16 Adding three diagonal lines to a hexagram yields the following solution:

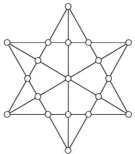

Figure S3.12

Chapter 4

4.1 See Figure S4.1 for one example based on the hint and another based on squares. Many others exist.

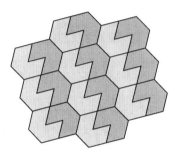 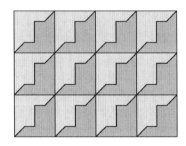

Figure S4.1

4.2 From Theorem 4.3.1 we have $bc = ac + ab$, which is equivalent to $2a = 2bc/(b+c)$.

4.3 The cubic in x is equivalent to $b^3 + a^3 = ab^2 + 2a^2b$. To derive this, multiply equation ii) in Theorem 4.3.1 by b ($b^3 = abc + a^2b$), equation ii) by a ($a^3 = ab^2 - a^2c$), equation iv) by a ($abc = a^2c + a^2b$), add and simplify.

Similarly, the cubic equation in y is equivalent to $c^3 + a^3 = 2ac^2 + a^2c$. To derive this, multiply equation i) by c ($c^3 = bc^2 + a^2c$), equation i) by a ($a^3 = ac^2 - abc$), equation iv) by c ($bc^2 = ac^2 + abc$), add and simplify.

4.4 (i) $(b + c - a)^2 = a^2 + b^2 + c^2 + 2(bc - ab - ac)$, but $a^2 + b^2 + c^2 = 7R^2$ from Theorem 4.4.4 and $bc - ab - ac = 0$ from Theorem 4.3.1(iv).

(ii) Theorem 4.4.1 and Lemma 4.4.3 yield
$$\frac{b^2}{a^2} + \frac{c^2}{b^2} + \frac{a^2}{c^2} = 4(\cos^2 A + \cos^2 B + \cos^2 C)$$
$$= 4[3 - (\sin^2 A + \sin^2 B + \sin^2 C)] = 4[3 - (7/4)] = 5.$$

(iii) $abc = 2R \sin A \cdot 2R \sin B \cdot 2R \sin C = 8R^3 \sqrt{7}/8 = \sqrt{7}R^3$.

4.5 Since the area of ABC is $K = ah_a/2$ we have $h_a = 2K/a$ (and similarly for h_b and h_c). Theorem 4.3.1 yields $h_a = 2K(1/a) = 2K(1/b + 1/c) = h_b + h_c$.

4.6 Let [ABC] denote the area of triangle ABC. Then $[ABC] = ab\sin C/2$ and $\sin C = c/(2R)$ so that $[ABC] = abc/(4R) = \sqrt{7}R^2/4$ as claimed.

4.7 Rotate $\triangle AXB$ in Figure 4.10.1 about X to $\triangle DXY$ as shown in Figure S4.2 and label the angles, where k denotes $k\pi/7$. Then $YD = AB$, and $AY = AX$ since $\triangle AXY$ is isosceles. Hence $AB + AX = YD + AY = AD$.

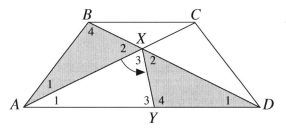

Figure S4.2

4.8 Let R denote the radius of the circle, P the midpoint of AB, Q the midpoint of the radius perpendicular to BC, and d the length of PQ, as shown in Figure S4.3.

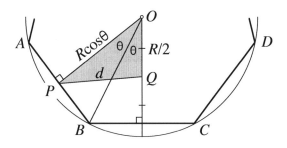

Figure S4.3

Applying the law of cosines to $\triangle OPQ$ yields

$$d^2 = (R/2)^2 + (R\cos\theta)^2 - 2(R/2)(R\cos\theta)\cos 2\theta$$
$$= R^2(1/4 + \cos^2\theta - \cos\theta\cos 2\theta).$$

Noting that $\sin 3\theta = \sin 4\theta$ (since $3\theta + 4\theta = \pi$), the triple angle sine formula yields

$$3\sin\theta - 4\sin^3\theta = \sin 3\theta = \sin 4\theta = 4\sin\theta\cos\theta\cos 2\theta,$$

hence

$$4\cos\theta\cos 2\theta = 3 - 4\sin^2\theta = -1 + 4\cos^2\theta$$

so that $\cos^2\theta - \cos\theta\cos 2\theta = 1/4$. Consequently $d^2 = R^2/2$ so that $d = R\sqrt{2}/2$, one-half the side of a square inscribed in the circle.

4.9 In Figure S4.4a we see that if $\angle APB = 120°$, then arc AB is one-third of the circumference of the circle, since the reflex angle $\angle AOB$ is twice $120°$.

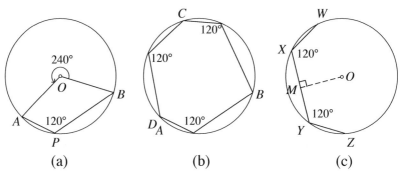

Figure S4.4

It follows that two of the 120° angles must be adjacent. Otherwise, as seen in Figure S4.4b, arcs AB, BC, and CD would account for the entire circumference, D would coincide with A, and the heptagon would collapse into a hexagon. Hence two 120° angles, X and Y in Figure S4.4c, are adjacent. Thus the portion $WXYZ$ of the heptagon is symmetric about OM the perpendicular bisector of XY, and consequently sides WX and YZ have equal length.

4.10 See Figure S4.5 and let a be the side length and b and c the lengths of the short and long diagonals, respectively.

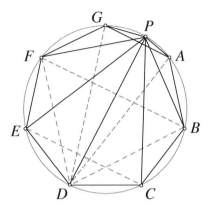

Figure S4.5

Applying Ptolemy's theorem to quadrilateral $PADG$ yields $aPD = c(PA + PG)$, to quadrilateral $PCDE$ yields $bPD = a(PC + PE)$, and to quadrilateral $PBDF$ yields $cPD = b(PB + PF)$. Hence

$$PA + PC + PE + PG = \left(\frac{a}{c} + \frac{b}{a}\right)PD$$
$$= \left(1 + \frac{c}{b}\right)PD = PB + PD + PF,$$

where in the middle step we used (4.6) to obtain $(a/c) + (b/a) = 1 + (c/b)$.

4.11 See Figure S4.6.

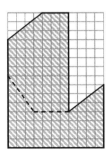 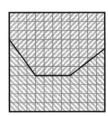

Figure S4.6

Chapter 5

5.1 See Figure S5.1 for two solutions.

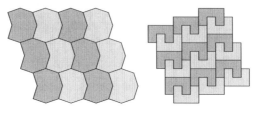

Figure S5.1

Many other solutions exist. For example we have the tiling with concave octagons on the left in Figure 1.7.4. In Figure S5.2 we have one based on another tiling in the Alhambra palace in Granada, Spain.

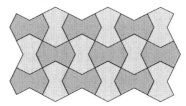

Figure S5.2

Note that each tile is the union of two irregular pentagons as in the tilings on the right in Figure 1.8.3 and in Figure 2.6.3a.

5.2 Rearrange the order of the eight sides of the octagon in Figure 5.7.1 to obtain the octagon in Figure S5.3 with the area A of the octagon unaltered. Thus $A = \left(x\sqrt{2} + y\right)^2 - x^2 = x^2 + y^2 + 2xy\sqrt{2}$.

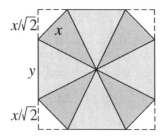

Figure S5.3

5.3 In Figure S5.4 we see the upper right quadrant of the octagon in Figure 5.7.2. If x denotes the radius of the circle, then both the diameter of the circle and the side length of the octagon equal $2x$.

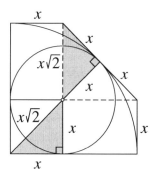

Figure S5.4

5.4 When $R = 1$ we have $a = \sqrt{2 - \sqrt{2}}$, $d_1 = \sqrt{2}$, $d_2 = \sqrt{2 + \sqrt{2}}$, and $d_3 = 2$. Hence the desired product is $a^2 d_1^2 d_2^2 d_3 = (2 - \sqrt{2})(2)(2 + \sqrt{2})(2) = 8$.

The general case is the following: If $A_1 A_2 \cdots A_n$ is a regular n-gon inscribed in a unit circle, $n \geq 3$, then $A_1 A_2 \cdot A_1 A_3 \cdot A_1 A_4 \cdot \cdots \cdot A_1 A_n = n$. For a proof using roots of unity see [Honsberger, 2003].

5.5 The octagon is equiangular, so we need only show that it is equilateral. Let the side of the square be $\sqrt{2}$ so that the radius of each arc is 1; and let x, y, and x denote the lengths of the three segments of each side of the square. Then $x + y = 1$ and $2x + y = \sqrt{2}$ so that $x = \sqrt{2} - 1$ and $y = 2 - \sqrt{2}$. The sides of the octagon measure y and $x\sqrt{2}$, and $x\sqrt{2} = 2 - \sqrt{2} = y$, thus the octagon is regular.

5.6 Assume the side lengths of the squares are $\sqrt{2}$ and 1. Drawing circular arcs with radius 1 at each vertex of the larger square reduces this construction to the one in the previous Challenge. See Figure S5.5.

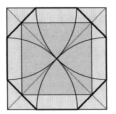

Figure S5.5

5.7 We deduce from the area of $ACEG$ that the radius of the circle is $\sqrt{5/2}$. If $s > t$ are the sides of the rectangle $BDFH$, then $s^2 + t^2 = 10$ and $st = 4$, so $(s+t)^2 = 18$ and $(s-t)^2 = 2$. Therefore $s + t = 3\sqrt{2}$ and $s - t = \sqrt{2}$, yielding $s = 2\sqrt{2}$ and $t = \sqrt{2}$. Without loss of generality, assume $BD = 2\sqrt{2}$ and $DF = \sqrt{2}$. See Figure S5.6.

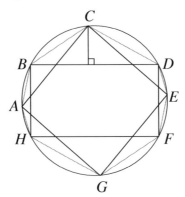

Figure S5.6

Let $[P]$ denote the area of a polygon P. By symmetry the area of the octagon can be expressed as $[BDFH] + 2[BCD] + 2[DEF]$. Note that $[BCD]$ is $\sqrt{2}$ times the distance from C to BD, which is maximized when C lies on the midpoint of arc BD. Similarly $[DEF]$ is $\sqrt{2}/2$ times the distance from E to DF, which is maximized when E lies on the midpoint of arc DF. Thus the area of the octagon is maximized when C lies on the midpoint of arc BD and E lies on the midpoint of arc DF (in which case $ACEG$ is indeed a square). In this case the distance from C to BD equals the radius of the circle minus half of DF, so $[BCD] = \sqrt{5} - 1$. Similarly $[DEF] = \sqrt{5}/2 - 1$, so the maximum area of the octagon is $3\sqrt{5}$.

5.8 Assume the side length a of each original octagon is 1. The fraction of the original octagon shaded gray in Figure 5.7.5a is $2 - \sqrt{2}$, or about 0.5858. The area K_o of the original octagon is $K_o = 2(\sqrt{2} + 1)$. The inradius $r_{\{8/2\}}$ of the gray inner octagon is $r_{\{8/2\}} = d_1/2 = \sqrt{2 + \sqrt{2}}/2$, and thus the area K_i of the inner gray octagon is $K_i = 8r_{\{8/2\}}^2(\sqrt{2} - 1) = 2\sqrt{2}$. Hence $K_i/K_o = \sqrt{2}/(\sqrt{2} + 1) = 2 - \sqrt{2}$.

The fraction of the original octagon shaded gray in Figure 5.7.5b is $3 - 2\sqrt{2}$, or about 0.1716. The inradius $r_{\{8/3\}}$ of the gray inner octagon is $r_{\{8/3\}} = 1/2$ so that its area K_i is $K_i = 2(\sqrt{2} - 1)$. Thus $K_i/K_o = (\sqrt{2} - 1)/(\sqrt{2} + 1) = 3 - 2\sqrt{2}$.

5.9 See Figure S5.7. Other solutions are possible.

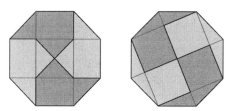

Figure S5.7

5.10 Each octagon is equilateral by the symmetry of the figures. We need only show that corresponding angles are equal. While the octagons are not equiangular, each octagon has two sets of four equal angles, i.e., four angles with measure α_1 alternating with four angles with measure β_1 in the octagon on the left, and similarly with α_2 and β_2 in the octagon on the right, as shown in Figure S5.8. We claim that, $\alpha_1 = \alpha_2$ and $\beta_1 = \beta_2$.

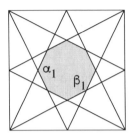 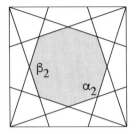

Figure S5.8

Using slopes of line segments and trigonometric identities yields $\tan\alpha_1 = \tan(2\arctan 2) = -4/3$ and $\tan\alpha_2 = -\tan(\arctan 3 - \arctan(1/3)) = -4/3$, thus $\alpha_1 = \alpha_2$. Similarly $\tan\beta_1 = -\tan(\arctan 2 - \arctan(1/2)) = -3/4$ and $\tan\beta_2 = \tan(2\arctan 3) = -3/4$, thus $\beta_1 = \beta_2$ and the octagons are similar.

If the octagons are inscribed in unit squares, the side length of the smaller is $\sqrt{5}/12$ and the side length of the larger is $\sqrt{10}/12$. Hence the ratio of their areas is $\left(\sqrt{10}/\sqrt{5}\right)^2 = 2$.

5.11 In Figure 5.7.8 the small gray octagon is also regular, so we have

$$\sqrt{2}+1 = \frac{m}{n} = \frac{n}{m-2n}$$

with n and $m-2n$ positive integers with $n < m$ and $m - 2n < n$, which contradicts the assumption that the fraction m/n was in lowest terms. Hence $\sqrt{2}+1$ (and $\sqrt{2}$) are irrational.

Chapter 6

6.1 In Figure S6.1 we see a portion of a regular nonagon with side length a and several diagonals. Applying the law of sines to $\triangle ACD$ yields $a/\sin 20° = d_2/\sin 120°$, and since $\sin 120° = \sqrt{3}/2$ we have $d_2 = (\sqrt{3}/2)a \csc 20°$.

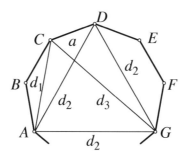

Figure S6.1

6.2 Applying van Schooten's theorem (Lemma 3.2.2) to the equilateral $\triangle ADG$ and vertex C in Figure S6.1 yields $d_3 = a + d_1$. The circumradius R equals 2/3 the altitude $(\sqrt{3}/2)d_2$ of $\triangle ADG$ so that $\sqrt{3}R = d_2$. The ratio r/R equals $\cos 20° = d_1/2a$, hence $2ar = Rd_1$. Lastly $\cot 20° = \cos 20° \csc 20° = d_1 d_2/\sqrt{3}a^2$ and hence $K = (3\sqrt{3}/4)d_1 d_2$.

6.3 Using the hint we have

$$4\sin 40° \sin 80° = 2\cos 40° - 2\cos 120°$$
$$= 2(1 - 2\sin^2 20°) + 1 = 3 - 4\sin^2 20°.$$

Hence

$$4\sin 20° \sin 40° \sin 80° = 3\sin 20° - 4\sin^3 20°$$
$$= \sin 60° = \sqrt{3}/2.$$

6.4 From (1.5) we have $d_1 = 2R\sin 40°$, $d_2 = \sqrt{3}R$, and $d_3 = 2R\sin 80°$. Since $a = 2R\sin 20°$ the preceding Challenge yields

$$ad_1 d_2 d_3 = 8\sqrt{3}R^4 \cdot (\sqrt{3}/8) = 3R^4.$$

6.5 Since $\angle MOB = 20°$ and $\angle BOC = 40°$, $\angle MOC = 60°$ and $\triangle MOC$ is equilateral, and shown in Figure S6.2.

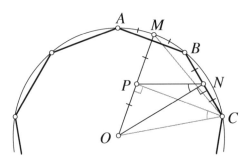

Figure S6.2

Thus $\angle CPO$ is a right angle and $\angle PCO = 30°$. But $\angle CNO$ is also a right angle, so that P and N lie on a semicircle with diameter OC. Hence $\angle PNO = \angle PCO = 30°$.

6.6 *Proof* 1. In Figure S6.3 AB and CD are sides of the decagon and AF and DF are two of the given diagonals. Draw diameters BE and CG through the center O. Since the decagon is regular, chords AB, CG, and DF are parallel, as are chords AF, BE, and CD. Hence the two shaded quadrilaterals are parallelograms, so that

$$DF = CH = HO + OC = AB + OC$$

as claimed [Honsberger, 2001].

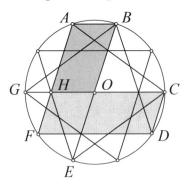

Figure S6.3

Proof 2. From Theorem 6.3.1 we have

$$d_2 = R\varphi = R\left(1 + \frac{1}{\varphi}\right) = R + a.$$

6.7 If $x^4 - x^2 = x^2 - x$ then $x^4 - 2x^2 + x = 0$, so that $x(x-1)(x^2+x-1) = 0$. Since $x > 0$ and $x \neq 1$, $x = (-1 + \sqrt{5})/2 = 1/\varphi$ as required.

6.8 Let C be the incenter of the triangle, and let A and B be the centers respectively of a circle through a vertex and a circle tangent to an adjacent side, as shown in Figure S6.4.

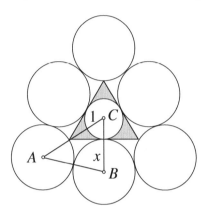

Figure S6.4

Let the inradius of the triangle equal 1, and let x be the radius of each of the surrounding circles. Then $\triangle ABC$ has sides $AB = 2x$, $BC = 1+x$, $AC = 2+x$, and $\angle C = 60°$. From the law of cosines we have $4x^2 = (1+x)^2 + (2+x)^2 - (1+x)(2+x)$, which simplifies to $x^2 - x - 1 = 0$. Hence $x = \varphi$, the golden ratio. The side of a regular decagon inscribed in a circle of radius φ equals $\varphi/\varphi = 1$, the inradius of the triangle.

6.9 Since $R/a = \varphi$ and $d_3/d_1 = \varphi$ it follows that $Rd_1 = ad_3$. Since $d_1 = R\sqrt{3-\varphi}$, $R = d_4/2$, and $K = 5R^2\sqrt{3-\varphi}/2$, we have $K = 5d_1d_4/4 = 5ar$, hence $d_1d_4 = 4ar$. Finally we have $d_1d_3 = R^2(3\varphi - \varphi^2) = \sqrt{5}R^2$ and $ad_2 = R^2$, hence $ad_1d_2d_3 = \sqrt{5}R^4$.

6.10 See Figure S6.5.

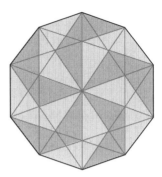

Figure S6.5

6.11 See Figure S6.6.

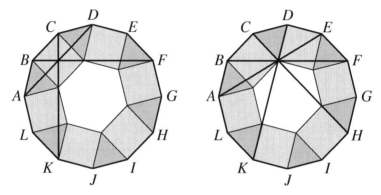

Figure S6.6

6.12 The square contains 16 congruent equilateral triangles and 16 congruent rhombi (four of which are the eight half-rhombi along the edges of the square). The dodecagon contains 12 of the equilateral triangles and 12 rhombi, thus the area of the dodecagon is 3/4 of the area 4 of the square, i.e., 3.

6.13 (i) Assume the dodecagon in Figure 6.10.5b has side length a and inradius r. The center portion of the dodecagon has area $2ar$, the area of the dodecagon is $6ar$, hence the fraction for the center portion is 1/3. By symmetry the other two portions also have area 1/3. (ii) Yes, see Figure S6.7a. Many other dissections are possible.

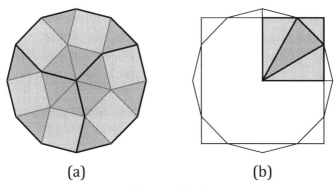

(a)　　　　　　　　(b)

Figure S6.7

6.14 Using the hint we have Figure S6.7b. Let K be area of the dodecagon, so that the area of the dark gray triangle is $K/12$. The large square also has area K from Example 6.5.2, so that area of the small square in two shades of gray is $K/4$, Hence the area of the triangle is $1/3$ of the area of the small square [Jobbings, 2011].

6.15 The hint yields imbedding $\triangle ABC$ in the 18-gon in Figure S6.8 (only its circumcircle and vertices are shown).

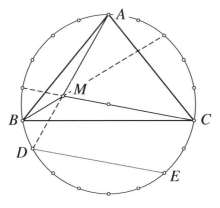

Figure S6.8

Extensions of segments AM, BM, and CM are concurrent diagonals (see [Prasolov, 2000] for a proof). Since DE is parallel to MC, $\angle AMC = \angle ADE = 70°$.

6.16 At least three of the vertices must have the same label (two vertices with each label only accounts for eight vertices). Vertices with the same label form a regular pentagon, and any three vertices of a regular pentagon form an isosceles triangle [Vaderlind et al., 2002].

6.17 See Figure S6.9.

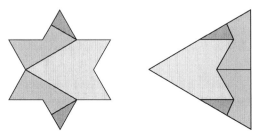

Figure S6.9

Chapter 7

7.1 The lower bound becomes 3/2 as illustrated in Figure S7.1.

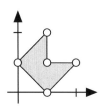

Figure S7.1

7.2 Suppose there exists an equilateral triangle with side s whose vertices are lattice points. By Heron's formula its area is $K = \sqrt{3}s^2/4$. Since the vertices are lattice points, s^2 is an integer and hence K is irrational. But by Pick's theorem K must be rational, a contradiction.

7.3 The answer to both questions is no. See the two concave lattice pentagons in Figure S7.2; the one on the left has area 4, the one on the right has area 4.5.

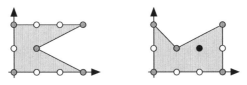

Figure S7.2

7.4 Let the simple rectilinear polygon have $2n$ sides (and vertices). Then $90a + 270b = 180(2n - 2)$ and $a + b = 2n$, from which it follows that that $a = b + 4$ (and $a = n + 2$, $b = n - 2$).

7.5 All four do. See Figure S7.3.

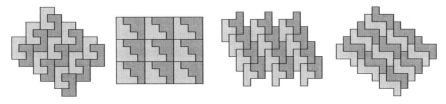

Figure S7.3

7.6 Triangulating the pentagon and employing (7.2) yields

$$K = \frac{1}{2}\left(\left\|\begin{matrix}1 & 0\\ \varphi & 1\end{matrix}\right\| + \left\|\begin{matrix}\varphi & 1\\ 1 & \varphi\end{matrix}\right\| + \left\|\begin{matrix}1 & \varphi\\ 0 & 1\end{matrix}\right\|\right)$$
$$= \frac{1}{2}[1 + (\varphi^2 - 1) + 1] = 1 + \frac{\varphi}{2}.$$

7.7 If P is equilateral, then it is also equiangular by Theorem 7.5.1. Assume P is equiangular, with its vertices and sides labeled as in Figure 7.5.3, but with A_{2n} and a_{2n} replaced by A_{2n-1} and a_{2n-1}. Proceeding as in Example 7.5.3, sides with odd subscripts have equal length as do sides with even subscripts, until we reach congruent triangles $\Delta A_1 A_2 A_3$ and $\Delta A_{2n-1} A_1 A_2$. Here we have $a_{2n-1} = a_2$, and hence all sides have equal length and P is equilateral.

7.8 If Q is equiangular, then it is also equilateral by Theorem 7.6.1. Assume Q is equilateral, with its vertices and sides labeled as in Figure 7.6.1, but with A_n and a_n replaced by

A_{2n-1} and a_{2n-1}. Then for the sides adjacent to A_2 we have $x_1 + x_2 = x_2 + x_3$ so that $x_1 = x_3$, and for the sides adjacent to A_3 we have $x_2 + x_3 = x_3 + x_4$ so that $x_2 = x_4$. Continuing in this fashion shows that the x_i with odd subscripts are equal, and the x_i with even subscripts are equal, until we reach the sides adjacent to A_1. Here we have $x_{2n-1} + x_1 = x_1 + x_2$ so that $x_{2n-1} = x_2$. Hence all the x_i are equal, and thus Q is equiangular as shown in the proof of Theorem 7.6.2.

7.9 (i) Yes. Alternate vertices of the $2n$-gon are the vertices of two congruent regular n-gons (whose sides are the d_1 diagonals of the $2n$-gon), both cyclic. The two n-gons are reflections of one another in each of the n lines connecting midpoints opposite sides, so the two circumcircles coincide.

(ii) No. See Figure S7.4.

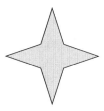

Figure S7.4

7.10 Yes, infinitely many. The answer to the question in the Hint is that both the triangle and square are tangential with inradius 2. From (7.4) every tangential polygon with inradius 2 is equable. There are many non-tangential equable polygons, for example, the rectilinear lattice hexagon with vertices (0,0), (6,0), (6,2), (3,2), (3,6), and (0,6) is equable with area and perimeter 24.

Chapter 8

8.1 For (i) and (ii) see Figure S8.1a and S8.1b, respectively.

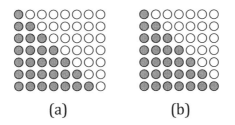

Figure S8.1

(iii) $P_{5,n} = T_n + 2T_{n-1} = n^2 + T_{n-1}$.

(iv) $P_{6,n} = \frac{1}{2}(4n^2 - 2n) = 2n^2 - n = n(2n-1) = T_{2n-1}$.

(v) $(2n-1)^2 - (n-1)^2 = 3n^2 - 2n = n^2 + 4T_{n-1}$
$$= T_n + 5T_{n-1} = P_{8,n}.$$

(vi) Use part (ii) with the second form for $P_{k,n}$ in (8.1).

8.2 For (i) and (ii) see Figure S8.2a and S8.2b, respectively.

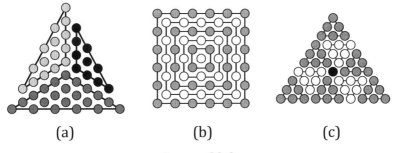

Figure S8.2

(iii) $n^3 - (n-1)^3 = 3n^2 - 3n + 1 = 1 + 6T_{n-1} = C_{6,n}$.

(iv) $C_{8,n} = 1 + 8T_{n-1} = 1 + 4n^2 - 4n = (2n-1)^2$

See Figure S8.3 for an illustration of (iv) (when $n = 3$).

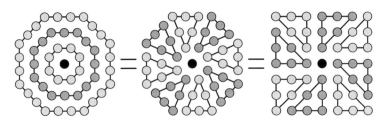

Figure S8.3

(v) $C_{9,n} = 1 + 9T_{n-1}$ and see Figure S8.2c for $1 + 9T_{n-1} = T_{3n-2}$ (when $n = 4$).

(vi) $C_{n,n+2} = 1 + nT_{n+1} = 1 + \frac{1}{2}n(n+1)(n+2) = 1 + (n+2)T_n = C_{n+2,n+1}$.

(vii) $C_{4,n} + 4T_{n-1} = 1 + 8T_{n-1} = 4n^2 - 4n + 1 = (2n-1)^2$.

(viii) $C_{6,n} + 3T_{n-1} = 1 + 9T_{n-1} = T_{3n-2}$ (see part (v) above).

See Figure S8.4ab for illustrations of (vii) and (viii), respectively, for $n = 4$.

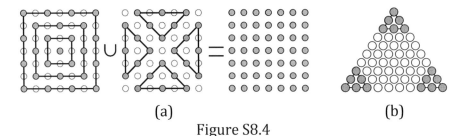

(a)   (b)

Figure S8.4

8.3 Using (8.2) and simplifying we have

$$P_{3k-4,n} = \frac{1}{2}[(3k-6)n^2 - (3k-8)n],$$

$$P_{k,2n-1} = \frac{1}{2}[(4k-8)n^2 - (6k-16)n + (2k-6)], \text{ and}$$

$$P_{k,n-1} = \frac{1}{2}[(k-2)n^2 - (3k-8)n + (2k-6)],$$

which proves the claim.

8.4 Using (8.2) we obtain

$(2k-1)P_{2k+1,n} + T_{k-2}$

$= \frac{1}{2}(2k-1)[(2k-1)n^2 - (2k-3)n] + \frac{1}{2}(k-2)(k-1),$

$= \frac{1}{2}[(2k-1)^2 n^2 - (2k-1)(2k-3)n + (k-1)(k-2)],$

$= \frac{1}{2}[(2k-1)n - (k-1)][(2k-1)n - (k-2)],$

$= T_{(2k-1)n-(k-1)}.$

8.5 (i) Since $C_{4,n} = 2n^2 - 2n + 1$ and $(2n-1)^2 = 4n^2 - 4n + 1$, $C_{4,n} = [(2n-1)^2 + 1]/2$. Hence
$(2n-1)^2 C_{4,n} = (2n-1)^2 [(2n-1)^2 + 1]/2 = T_{(2n-1)^2}.$

(ii) Since $H_n = 1 + 3n(n-1)$ and $S_n = 1 + 6n(n-1)$, $H_n = (S_n + 1)/2$. Hence $H_n S_n = S_n(S_n + 1)/2 = T_{S_n}.$

(iii) From the hint, the generalization is $C_{k,n} C_{2k,n} = T_{C_{2k,n}}.$ Since $C_{2k,n} = 2C_{k,n} - 1$ we have $C_{k,n} = (C_{2k,n} + 1)/2$ so that $C_{k,n} C_{2k,n} = C_{2k,n}(C_{2k,n} + 1)/2 = T_{C_{2k,n}}.$

8.6 Following the hint, we have $18n(n-1) + 5 = 3m^2 + 2$. Note thar $18n(n-1) + 5 = (3n-1)^2 + (3n-2)^2$ and $3m^2 + 2 = (m-1)^2 + m^2 + (m+1)^2$ [Gardner, 1988].

8.7 From (8.2) the nth k-gonal number $P_{k,n}$ is

$$\frac{n}{2}[(k-2)n - (k-4)] = \frac{n}{2}[(k-2)(n-1) + 2].$$

Assume $n \geq 3$ and $k \geq 3$. When n is even $P_{k,n}$ factors as $(n/2) \cdot [(k-2)n - (k-4)]$ with both factors greater than 1. When n is odd, $n-1$ is even and $P_{k,n}$ factors as $n \cdot ([(k-2)(n-1) + 2]/2)$ with both factors greater than 1.

8.8 For a prime p such that $2^p - 1$ is also prime, the perfect number $N_p = T_{2^p - 1}$. The nth nonagonal number $C_{9,n} = T_{3n-2}$ from Challenge 8.2. Since odd powers of 2 are 1 less than a multiple of 3, for an odd prime p, $2^p + 1 = 3n$ so that $2^p - 1 = 3n - 2$. Thus $N_p = T_{2^p - 1} = T_{3n-2} = C_{9,(2^p+1)/3}$, e.g., $28 = N_3 = C_{9,3}$ and $496 = N_5 = C_{9,11}$.

8.9 Yes, infinitely many. The equation $H_n = S_m$ is equivalent to $n(n-1) = 2m(m-1)$, which in turn can be written as $n^2 + (n-1)^2 = (2m-1)^2$. This is the Diophantine equation in Example 8.3.3 (with m replaced by $2m-1$) that had infinitely many solutions. So the solutions (n, m) to $H_n = S_m$ are (1,1), (4,3), (21,15), (120,85), etc., and the corresponding "star hexes" are 1, 37, 1261, 42841, etc.

Credits and Permissions

Cover: Source: Wikimedia Commons, author Tomruen, licensed under the Creative Commons Attribution-Share Alike 4.0 International license (https://creativecommons.org/licenses/by-sa/4.0/deed.en).

Portrait of Gauss in Section 1.5: Source: Wikimedia Commons, public domain.

Portrait of Wantzel in Section 1.5: Source: MacTutor.

Figure 1.7.1a: Source: Wikimedia Commons, author Mbellaccini, licensed under the Creative Commons Attribution-Share Alike 4.0 International license (https://creativecommons.org/licenses/by-sa/4.0/deed.en).

Figure 1.7.1b: Source: Wikimedia Commons, author Anneke Bart, licensed under the Creative Commons Attribution-Share Alike 3.0 Unported license (https://creativecommons.org/licenses/by-sa/3.0/deed.en).

Figure 1.8.1a: Source: Wikimedia Commons, public domain.

Figure 2.1.2a: Source: Wikimedia Commons, author James St. John, licensed under the Creative Commons Attribution 2.0 Generic license (https://creativecommons.org/licenses/by/2.0/deed.en).

Figure 2.1.2b: Source: Wikimedia Commons, author Rob Lavinsky, iRocks.com, licensed under the Creative Commons Attribution-Share Alike 3.0 Unported license (https://creativecommons.org/licenses/by-sa/3.0/deed.en).

Figure 2.1.2c: Source: Wikimedia Commons, author Aravind Sivaraj, licensed under the Creative Commons Attribution-Share Alike 3.0 Unported license (https://creativecommons.org/licenses/by-sa/3.0/deed.en).

Figure 2.2.4: Source: Wikimedia Commons, author Diderot, licensed under the Creative Commons CC0 1.0 Universal Public Domain Dedication (https://creativecommons.org/publicdomain/zero/1.0/deed.en).

Figure 2.3.4: Source for all nine images: Wikimedia Commons, public domain.

Figure 2.4.3a: Source: Wikimedia Commons, public domain.

Figure 2.6.3a: Source: Wikimedia Commons, author David Eppstein, licensed under the Creative Commons CC0 1.0 Universal Public Domain Dedication (https://creativecommons.org/publicdomain/zero/1.0/deed.en).

Figure 2.6.8b: © Mathematical Association of America, 1985. All rights reserved.

Figures 2.7.3abc: Source: Wikimedia Commons, public domain.

Figures 2.7.5ab: Source: Wikimedia Commons, public domain.

Figure 2.8.3ab: Source: Wikimedia Commons, public domain.

Figure 2.10.1: Source: Wikimedia Commons, public domain.

Figure 2.10.2a: Source: Wikimedia Commons, author Smecucci, licensed under the Creative Commons Attribution-Share Alike 3.0 Unported license (https://creative commons.org/licenses/by-sa/3.0/deed.en).

Figure 2.10.2b: Source: Wikimedia Commons, author Claudio Stanco, licensed under the Creative Commons Attribution-Share Alike 3.0 Unported license (https://creativecommons.org/licenses/by-sa/3.0/deed.en).

Figure 2.10.3: Source: Wikimedia Commons, public domain.

Figure 2.11.4a: Source: Sam Loyd's Cyclopedia of Puzzles.

Figure 2.11.5: Source: Sam Loyd's Cyclopedia of Puzzles.

Figure 3.1.5a: Source: Wikimedia Commons, author AJC1, licensed under the Creative Commons Attribution-Share Alike 2.0 Generic license (https://creativecommons.org/licenses/by-sa/2.0/deed.en).

Figure 3.1.5b: Source: Wikimedia Commons, author Alexey Kljatov, licensed under the Creative Commons Attribution-Share Alike 4,0 International license (https://creativecommons.org/licenses/by-sa/4.0/deed.en).

Figure 3.2.3: Source for leftmost three images in the first row: Wikimedia Commons, licensed under the Creative Commons CC0 1.0 Universal Public Domain Dedication (https://creativecommons.org/publicdomain/zero/1.0/deed.en); source for the other images: Wikimedia Commons, public domain.

Figure 3.2.7a: Source: Wikimedia Commons, author Mathsci, licensed under GNU Free Documentation License Version 1.2 or any later version published and the Creative Commons Attribution-Share Alike 3.0 Unported license (https://creativecommons.org/licenses/by-sa/3.0/deed.en).

Figure 3.2.11a: Source: Wikimedia Commons, public domain.

Figure 3.2.11b: Source: Wikimedia Commons, licensed under the Creative Commons CC0 1.0 Universal Public Domain Dedication (https://creativecommons.org/publicdomain/zero/1.0/deed.en). Credit: Purchase, Rogers Fund; Thomas A. D. Ettinghausen and Elizabeth S. Ettinghausen Gifts, in memory of Richard Ettinghausen; and Steven Kossak, Mrs. Charles Wrightsman, and Richard S. Perkins Gifts, 1995.

Figure 3.2.11c: Source: Wikimedia Commons, licensed under the Creative Commons CC0 1.0 Universal Public Domain Dedication (https://creativecommons.org/publicdomain/zero/1.0/deed.en). Credit: Bequest of Carolyn Fiske MacGregor, 1953, in memory of her grandmother, Caroline Brooks Gould.

Figure 3.4.1: Source: Wikimedia Commons, author Andreu Ledoux, licensed under the Creative Commons Attribution-Share Alike 3.0 Unported license (https://creativecommons.org/licenses/by-sa/3.0/deed.en).

Figure 3.8.2a: Source: Wikimedia Commons, licensed under the Creative Commons CC0 1.0 Universal Public Domain Dedication (https://creativecommons.org/publicdomain/zero/1.0/deed.en). Credit: Severance and Greta Millikin Purchase Fund.

Figures 3.8.2bc: Source: Wikimedia Commons, public domain.

Figure 3.9.7: Source: Wikimedia Commons, public domain.

Figure 3.10.1a: Source: Wikimedia Commons, public domain.

Figure 3.10.1b: Source: Wikimedia Commons, licensed under the Creative Commons CC0 1.0 Universal Public Domain Dedication (https://creativecommons.org/publicdomain/zero/1.0/deed.en). Credit: Het Utrechts Archief.

Figure 3.10.2a: Source: Wikimedia Commons, author Xephro, licensed under the Creative Commons CC0 1.0 Universal Public Domain Dedication (https://creativecommons.org/publicdomain/zero/1.0/deed.en).

Figure 3.10.2b: Source: Wikimedia Commons, public domain.

Figure 3.10.3: Source: Wikimedia Commons, author Frédéric Neupont, licensed under the Creative Commons CC0 1.0 Universal Public Domain Dedication (https://creativecommons.org/publicdomain/zero/1.0/deed.en).

Figure 4.1.1: Source: Wikimedia Commons, licensed under the Creative Commons Attribution-Share Alike 2.0 France license (https://creativecommons.org/licenses/by-sa/2.0/fr/deed.en).

Figure 4.5.1b: Source: Bildnachwies: SLUB/Deutsche Fotothek/DDZ, public domain.

Figure 4.9.1a: Source: Wikimedia Commons, author Rabanus Flavus, licensed under the Creative Commons Attribution-Share Alike 3.0 Unported license (https://creativecommons.org/licenses/by-sa/3.0/deed.en).

Figure 4.9.1b: Source: Wikimedia Commons, author Rabanus Flavus, licensed under GNU Free Documentation License Version 1.2 or any later version published and Creative Commons Attribution-Share Alike 3.0 Unported license (https://creativecommons.org/licenses/by-sa/3.0/deed.en).

Figure 4.9.2a: Source: Wikimedia Commons, author João Lopes, licensed under the Creative Commons Attribution-Share Alike 3.0 Unported license (https://creativecommons.org/licenses/ by-sa/3.0/deed.en).

Figure 4.9.2b: Source: Wikimedia Commons, author Hispalois, licensed under GNU Free Documentation License Version 1.2 or any later version published and Creative Commons Attribution-Share Alike 4.0 International license (https://creativecommons.org/licenses/by-sa/4.0/deed.en).

Figure 4.9.3: © Mathematical Association of America, 2009. All rights reserved.

Figure 4.10.2a: Source: Sam Loyd's Cyclopedia of Puzzles.

Figure 5.1.2: Source for the leftmost image in the first row: Wikimedia Commons, licensed under the Creative Commons CC0 1.0 Universal Public Domain Dedication (https://creativecommons.org/publicdomain/zero/1.0/deed.en). Source for the left-most image in the second row: Wikimedia Commons, author Kristoffer Gustafsson, licensed under the Creative Commons Attribution-Share Alike 4.0 International license (https://creativecommons.org/licenses/by-sa/4.0/deed.en). Source for the others: Wikimedia Commons, public domain.

Figure 5.2.2a: Source: Wikimedia Commons, courtesy of Heritage Auctions, licensed under the Creative Commons Attribution 4.0 International license (https://creativecommons.org/licenses/by/4.0/deed.en).

Figure 5.2.2b: Source: Wikimedia Commons, author kevin-mcgill, licensed under the Creative Commons Attribution-Share Alike 2.0 Generic license (https://creativecommons.org/licenses/by-sa/2.0/deed.en).

Figure 5.2.2c: Source: Wikimedia Commons, licensed under the Creative Commons CC0 1.0 Universal Public Domain Dedication (https://creativecommons.org/publicdomain/zero/1.0/deed.en).

Figures 5.2.2d: Source: Wikimedia Commons, public domain.

Figure 5.2.8: Source: Wikimedia Commons, author Bjoertvedt, licensed under the Creative Commons Attribution-Share Alike 4.0 International license (https://creativecommons.org/licenses/by-sa/4.0/deed.en).

Figure 5.4.3: Source: Wikimedia Commons, author Serge Melki, licensed under the Creative Commons Attribution 2.0 Generic license (https://creativecommons.org/licenses/by/2.0/deed.en).

Figures 5.5.1b: Source: Wikimedia Commons, public domain.

Figures 5.6.1ab: Source: Wikimedia Commons, public domain.

Figures 5.6.2ab: Source: Wikimedia Commons, public domain.

Figures 5.6.3ab: Source: Wikimedia Commons, public domain.

Figure 5.6.4: Source: Wikimedia Commons, public domain.

Figure 5.6.5: Source: Wikimedia Commons, public domain.

Figures 6.2.7ab: Source: Wikimedia Commons, public domain.

Figures 6.3.6ab: Source: Wikimedia Commons, public domain.

Figure 6.4.1a: Source: Wikimedia Commons, public domain.

Figures 6.4.2ab: Source: Wikimedia Commons, public domain.

Figure 6.5.8: Source: Wikimedia Commons, author Jebulon, licensed under the Creative Commons CC0 1.0 Universal Public Domain Dedication (https://creativecommons.org/publicdomain/zero/1.0/deed.en).

Figure 6.5.9ab: Source: Wikimedia Commons, Scan: Retroplum, under Creative Commons Attribution-Share Alike 3.0 Unported License (https://creativecommons.org/licenses/by-sa/3.0/deed. en).

Figure 6.6.1a: Source: Wikimedia Commons, public domain.

Figure 6.6.2: Source: Wikimedia Commons, author Gnt305, licensed under the Creative Commons Attribution-Share Alike 4.0 International license (https://creativecommons.org/licenses/by-sa/4.0/deed.en).

Figure 6.7.1: Source: Wikimedia Commons, public domain.

Figure 6.7.2: Jiu zhang suan shu, juan 1. Liu Hui (ca. 3rd century), Li Chunfeng (602-670) et al. Shanghai: [publisher not identified], Qing Guangxu 16 [1890]. 九章算術, 卷一。 劉徽注, 李淳風等注釋。上海: [不明], 清光緒庚寅 [16 年, 1890]. Smith Chinese C-9. folio 22. Smith-Plimpton East Asian Books, Art, and Manuscript collection. Rare Book and Manuscript Library, Columbia University Libraries.

Portrait of Georg Pick in Section 7.2: Source: Wikimedia Commons, public domain.

Figure 7.3.2: Source: Wikimedia Commons, public domain.

Figures 7.7.2ab: Source: Wikimedia Commons, public domain.

Figure 8.2.3: © Mathematical Association of America, 2014. All rights reserved.

Figure 8.4.1c: Source: Wikimedia Commons, public domain.

Figure 8.4.2: Source for all four images: Wikimedia Commons, public domain.

Figure S2.8: Source: Sam Loyd's Cyclopedia of Puzzles.

Bibliography

Alexandrov, A. D. 2005. *Convex Polyhedra*. Springer, Berlin. MR2127379

Alsina, C. and Nelsen, R. B., 2009. *When Less is More*. Mathematical Association of America, Washington. MR 2498836

Alsina, C. and Nelsen, R. B., 2010. *Charming Proofs*. Mathematical Association of America, Washington. MR2675936

Alsina, C. and Nelsen, R. B., 2020. *A Cornucopia of Quadrilaterals*. Mathematical Association of America, Washington. MR 4286138

Arkinstall, J. R., 1980. *Minimal requirements for Minkowski's theorem in the plane I*, Bull. Austral. Math. Soc. **22**, 259–274. MR0598699

Augros, M., 2012. *Rediscovering Pascal's mystic hexagon*. Coll. Math. J. **43**, 194-202. MR 2916484

Ball, W. W. R. and Coxeter, H. S. M., 1974. *Mathematical Recreations & Essays, 12th Ed.*, University of Toronto Press, Toronto. MR0351741

Bankoff, L. and Garfunkel, J., 1973. *The heptagonal triangle*, Math. Magazine **46**, 7-19. MR1572030

Bell, E. T., 1951. *Mathematics, Queen and Servant of Science*. McGraw-Hill, New York. MR0960277

Beyer, W. A., Louck, J. D., Zellberger, D., 1996. *A generalization of a curiosity that Feynman remembered all his life*. Math. Magazine **69**, 43-44. MR1573132

Bliss, N., Fulan, B., Lovett, S., Sommars, J., 2013. *Strong divisibility, cyclotomic polynomials, and iterated polynomials*. Amer. Math. Monthly **120**, 519-536. MR3063117

Bogomolny, A., 2018. https://www.cut-the-knot.org/triangle/MidpointsInHexagon2.shtml.

Bolt, B., Eggleton, R., and Gilks, J., 1991. *The magic hexagram*. Math. Gazette **75**, 140-142.

Boyer, C. B., 1968. *A History of Mathematics*. John Wiley & Sons, New York. MR0234791

Brunton, J. K., 1961. *Polygonal knots*, Math. Gazette **45**, 299-302.

Conway, J. H. and Guy, R. K., 1996. *The Book of Numbers*, Springer, New York. MR1411676

Coxeter, H. S. M. and Greitzer, S. L., 1967. *Geometry Revisited*, Random House, New York. MR3155265

David, G., and Tomei, C. *The problem of the calissons*. Amer. Math. Monthly **96**, 429-431. MR0994034

de la Hoz, R., 1996. *La Proporción Cordobesa*, Actas de las VII Jornadas Andaluzas de Educación Matemática Thales. Ed. Servicio de Publicaciones de la Universidad de Córdoba, 65-74.

de Villiers, M. 2007. *A hexagon result and its generalization via proof*. Mont. Math. Enth. **4**, 188-192.

de Villiers, M. 2011a. *Proof without words: Parahexagon-parallelogram area ratio*, Learning & Teaching Mathematics **10**, 23.

de Villiers, M., 2011b. *Equi-angled cyclic and equilateral circumscribed polygons*. Math. Gazette **95**, 102-106.

de Villiers, M. 2016. *Generalising some geometrical theorems and objects*. Learning & Teaching Mathematics **21**, 17-21.

Demar, R. F., 1975. *A simple approach to isoperimetric problems in the plane*. Math. Magazine **48**, 1-12.

Demir, H., 1966. *Maximum area of a region bounded by a closed polygon with given sides*. Math. Magazine **39**, 228-231. MR1571631

DeTemple, D. W., 1999. *Carlyle circles and the Lemoine simplicity of polygon constructions*. Amer. Math. Monthly **98**, 97-108. MR1089454

Devadoss, S. L. and O'Rourke, J., 2011. *Discrete and Computational Geometry*, Princeton University Press, Princeton and Oxford. MR2790764